BUILDING TOMORROW:
INNOVATION IN CONSTRUCTION AND ENGINEERING

Building Tomorrow: Innovation in Construction and Engineering

Edited by

ANDRÉ MANSEAU
Université de Québec à l'Outaouais, Canada

and

ROB SHIELDS
University of Alberta, Canada

ASHGATE

© André Manseau and Rob Shields 2005

All rights reserved. No part of this publication may be reproduced, stored in a retrieval system or transmitted in any form or by any means, electronic, mechanical, photocopying, recording or otherwise without the prior permission of the publisher.

André Manseau and Rob Shields have asserted their right under the Copyright, Designs and Patents Act, 1988, to be identified as the editors of this work.

Published by
Ashgate Publishing Limited
Gower House
Croft Road
Aldershot
Hants GU11 3HR
England

Ashgate Publishing Company
Suite 420
101 Cherry Street
Burlington, VT 05401-4405
USA

Ashgate website: http://www.ashgate.com

British Library Cataloguing in Publication Data
Building tomorrow : innovation in construction and
    engineering
    1.Construction industry - Technological innovations
    2.Construction industry - Management 3.Building -
    Technological innovations
    I.Manseau, André II.Shields, Rob, 1961-
    690

Library of Congress Cataloging-in-Publication Data
Building tomorrow : innovation in construction and engineering / edited by André Manseau and Rob Shields.
        p. cm.
    Includes bibliographical references and index.
    ISBN 0-7546-4378-6
    1. Building--Technological innovations. I. Manseau, André. II. Shields, Rob, 1961-

TH155.B785 2005
690--dc22
2005001841

ISBN 0 7546 4378 6

Printed and bound by Athenaeum Press, Ltd.,
Gateshead, Tyne & Wear.

Contents

List of Tables and Figures   *vii*
List of Contributors   *ix*
Preface by Sherif Barakat   *xi*

Introduction
    André Manseau and Rob Shields   1

1   A Survey of the Construction Innovation Literature
    Rob Shields   5

2   Construction – A Changing Industry Challenging Current Innovation Models
    André Manseau   23

3   Redefining Innovation
    André Manseau   43

4   Measuring Innovation in Construction
    Frances Anderson   57

5   Managing Complex Connective Processes: Innovation Broking
    Graham Winch   81

6   Understanding Risks and Shaping Large Engineering Projects
    Roger Miller and Serghei Floricel   101

7   The Government Role – From Market Failures to Social Capital
    André Manseau and Elisabeth Campagnac   123

8   Construction Industry Paradigms: The Final Frontier
    Andre Dorée and Frens Pries   139

9   Skills and Occupational Cultures
    Rob Shields   157

Conclusion: A Roundtable on Construction Innovation – How to Make it Work?
    André Manseau and Rob Shields with Contributors and Invited Experts   175

Index   *181*

List of Tables and Figures

Table 1.1	Impediments to innovation	18
Table 2.1	Key stakeholders in construction	29
Table 2.2	Three levels of economic analysis – micro, meso and macro	31
Table 2.3	Major changes occurring in the construction industry	35
Table 3.1	Sources of competitive advantage for contractors	52
Table 4.1	Industries classified in the construction sector by the North American Industrial Classification System 1997	58
Table 4.2	List of innovative activities in the 3rd CIS Innovation Survey	63
Table 4.3	The application of the systemic approach framework to the measurement of innovation in private sector firms	67
Table 4.4	Top four business strategies for organizational success and the top five characteristics of the competitive environment for the Queensland road and bridge industry	76
Table 6.1	Examples of turbulent events affecting projects	106
Table 6.2	Devices used in building strategic systems	111
Table 7.1	Type of construction goods	125
Table 7.2	Types of government intervention at local, national and international levels for supporting innovation and improvements in construction	135
Figure 2.1	Construction share to the economy	25
Figure 2.2	Productivity trends in terms of value-added per total revenue	26
Figure 2.3	Labour productivity trends	27
Figure 2.4	Productivity comparison	28
Figure 2.5	Key agents, major types of interactions and framework conditions in the construction sector	32
Figure 4.1	Actors in the construction cluster	59
Figure 4.2	Percentage of firms with innovation expenditures, by type of expenditure	64
Figure 4.3	Percentage of product innovators in building products and construction equipment, machinery and tool industries, by selected manufacturing industries 1997-1999	65
Figure 4.4	Percentage of businesses indicating use or planned use of advanced technologies within two years, all construction industries	69
Figure 4.5	Percentage of businesses indicating current use or planned use of practices within two years, all construction industries	70
Figure 4.6	Actors involved in the transmission of new products	73
Figure 4.7	Share of co-operation partners of innovators	75

Figure 5.1	Construction as a complex systems industry	89
Figure 5.2	The top-down and bottom-up modes of construction innovation	94
Figure 5.3	The diffusion of practice in construction	96
Figure 5.4	Practices and the diffusion of innovations	96
Figure 6.1	Projects differ substantially	103
Figure 6.2	Major risks in large engineering projects	104
Figure 6.3	Dynamic evolution of risks	105
Figure 6.4	Strategies to cope with risks	108
Figure 6.5	Shaping efforts	118
Figure 8.1	Parties and types of innovations	142
Figure 8.2	Sources of innovation	143
Figure 8.3	Average number of innovations per year	144
Figure 8.4	Stages in the innovation process	148
Figure 8.5	Distribution of advertisements: Construction and other sectors	151
Figure 8.6	Distribution of disciplines	151
Figure 8.7	Distribution of advertisements	152
Figure 8.8	Distribution of advertisements by qualification	152
Figure 8.9	Distribution of advertisements by firm size	153

List of Contributors

Frances Anderson is Senior Advisor on Indicators and Standards for Science and Technology Statistics, in the Science, Innovation and Electronic Information Division at Statistics Canada. She received her Ph.D. from the University of Montreal in History and Socio-Politics of Sciences.
Email: Frances.Anderson@statcan.ca

Sherif Barakat is CIB President 2001-2004 and Director General, Institute for Research in Construction, National Research Council Canada, Ottawa. He has a distinguished career in construction research. He holds a Ph.D. in Mechanical Engineering.

Elisabeth Campagnac, lawyer and sociologist, is Research Director at the Ecole Nationale des Ponts et Chaussées (Laboratoire Techniques, Territoires et Société), lecturer at Université de Paris I Panthéon-Sorbonne and visiting lecturer at University College London. She is a member of the editorial boards of *Les Annales des Ponts et Chaussées* and *Construction Management and Economics* and holds a doctorate in urban studies from Institut d'urbanisme de Grenoble.
Email: campagnac@mail.enpc.fr

Andre Dorée is Professor of Market and Organisations in the Construction Industry at the University of Twente in the Netherlands.
Email: A.G.Doree@ctw.utwente.nl

Serghei Floricel is Professor of Project Management at the Université du Québec Montreal. He has a Ph.D. in Administration, an MBA, and a Bachelor of Mechanical Engineering. His research focuses on the management of innovation and the management of large-scale projects. He is co-author of *The Strategic Management of Large Engineering Projects* (MIT) and currently Research Director for the SSHRC strategic grant 'Managing Innovation in the New Economy (MINE)'.
E-mail: floricel.serghei@uqam.ca

André Manseau is Director, University-Milieu Liaison Office, Université du Québec en Outaouais, on secondment from the National Research Council Canada. He holds a Ph.D. in Management (Université de Québec à Montréal). He co-authored *Canada's National System of Innovation* (McGill-Queens), with J. Niosi and B. Godin and co-edited *Innovation in Construction – An International Review of Public Policies,* (Spon) with George Seaden.
Email: andre.manseau@uqo.ca

Roger Miller is Hydro-Quebec/NSERC/SSHRC Chair in Technology Management at Université de Québec à Montréal. He is co-editor with Donald Lessard of The Strategic Management of Large Engineering Projects: Shaping institutions, risks, and governance (MIT).
Email: miller.roger@uqam.ca

Frens Pries is Managing Director of Balance & Result Management Consultants (The Netherlands) and holds a Ph.D. in Management.
Email: f.pries@balance-result.nl

Rob Shields founded and edits the journal *Space and Culture* and holds the Henry Marshall Tory Chair and Professorship in Sociology / Art and Design at the University of Alberta, Edmonton. He trained in Architecture and in Urban and Regional Studies.
Email: rshields@ualberta.ca

Graham Winch is Professor of Project Management in Manchester Business School, University of Manchester, and Director of the Centre for Research in the Management of Projects. He is the author of *Managing Production : Engineering Change and Stability* (Oxford)*, Innovation and Management Control* (Cambridge), and most recently, *Managing Construction Projects, an Information Processing Approach* (Blackwell, 2002).
Email: graham.winch@mbs.ac.uk

Preface

Innovation is widely recognized as a key factor for developing competitive advantages in any economic activity. It is also recognized that the construction industry is a crucial sector for achieving wealth and quality of life. However, a number of national and international panels identified that construction remains relatively less innovative than many other sectors although many stakeholders dedicate significant efforts for enhancing this sector and its impact to the overall economy. Innovation in construction has been a rapidly growing field of study in the last decades. A number of studies from different perspectives have been conducted but it is still generally difficult to have a comprehensive view offered in one text.

This book responds to this very challenge and will no doubt be of a great interest to a large audience of researchers, managers and policy makers involved in construction. I also believe the book addresses a crucial need for a handbook on managing innovation in construction. With the contribution of leading experts from Canada, the United Kingdom, France, the Netherlands and Australia this work provides a wide variety of perspectives. These include project management, firm management, labour training, industry statistics and measurement, culture, macro-economics and public policy.

The International Council for Research and Innovation in Building and Construction (CIB) was established in 1953 to stimulate and facilitate international cooperation and information exchange between governmental research institutes in the building and construction sector, with an emphasis on those institutes engaged in technical fields of research. The first approach was essentially a Technology Push model, based on the assumption that technology was the only key factor for initiating innovation.

CIB has since developed. Its approach to innovation has become much more complex, based on its world-wide network of over 5000 experts from about 500 member organizations active in the research community, in industry and in education. Although within the CIB program considerable attention is still given to technical topics, there are now also activities focused on such topics as organization and management, economics of building, legal and procurement practices, architecture, urban planning and human aspects. CIB is the world's foremost platform for international cooperation and information exchange in the area of building and construction research and innovation.

Managing innovation and changes in construction now involves a number of areas. There are still many questions that remain unanswered, but this book will definitely help in drawing out the many approaches for understanding the issue.

I had the opportunity to participate in the discussion which gave rise to the concluding chapter along with all authors and other experts, as we were linked via a telephone conference. The discussion was very informative – and somehow surprising – destabilizing former ways of thinking. We all came to realize that one of main problems of innovation in construction is not technology or poorly performing technologies. The most fundamental recent driver of innovation has been the change in client behavior with new contractual arrangements requiring long-term partnership and a life cycle perspective for buildings and facilities. Customers, and particularly governments that are still large clients, have introduced new forms of procurements in many countries. National visions or action plans, sponsored by governments, have had a significant impact.

However, construction professionals appear to have difficulty in addressing that need, and even more difficulties in developing new opportunities in this new context. Only a few large international construction firms have been successful and developed significant competitive advantage by investing in "soft-and-wise" issues such as strategic positioning, social inter-plays between players and project governance.

I am very pleased to introduce this book which goes beyond a discussion of technology to managing innovation and change in the construction industry. It promotes a better understanding of the business, competition, culture, policies and environmental changes that are required to advance innovation in the construction industry.

Sherif Barakat, Ph.D.
President CIB 2001-2004 and
Director General, Institute for Research in Construction
National Research Council Canada

Introduction

André Manseau and Rob Shields

This book fills a gap in the literature by bringing economic, social and construction and engineering management perspectives together. It marks the developing interest in innovation and the performance of construction and related industries. Management specialists not trained in civil engineering have entered the field and have remarked that key studies are difficult to locate. The current volume offers a unique survey of key issues and approaches. This book provides one of the first and comprehensive overviews of the rapidly growing new field of studies of 'Innovation in Construction' and its impact on economic, management and policy models and applications.

Construction and engineering are leaving behind the stereotypes they have often been viewed through. Finding its own voice and specificity, the use of over-generalized models of innovation, drawn from manufacturing industries, has been challenged. We have brought together a team of leading researchers from different countries to present different aspects of innovation in construction and related industries.

Contemporary research has its roots in the same factors related to innovation as those identified by writers such as Marion Bowley and Steven Groak. In Chapter 1 we present the evolution of ways of conceiving and approaching innovation in construction, engineering and related industries. Historically, innovation has been approached as a qualitative aspect of change. Typologies were developed to model innovation, but these approaches suffered from the use of overlapping categories and a retrospective approach that was not conducive to rigorous testing. These include the question of barriers to innovation, the diffusion process, the distinctive division and rivalry between builders, designers and users of buildings and research bias in favour of major changes as opposed to research on minor and incremental innovations.

Chapter 2 continues with a more detailed survey of attempts to model innovation since the Second World War. Science-based, 'technology push' models capture innovation in medicine or chemical industries, but proved inappropriate to the more market-oriented development of consumer products. Approaches centred on networks of firms, such as the OECD's 'Oslo' model of the 1990s, were specifically modeled on the clusters of firms in concentrated in high tech locations such as Silicon Valley. Challenges, however, have come from organizational approaches based on competitive change in the service sector and the behaviour of assembly firms at the heart of large, complex product systems (CoPS). Construction shares some characteristics with all of these approaches, depending on the state of economic cycles (for example, boom periods of building up new infrastructures in an expanding economy or maintaining existing buildings once market needs have been met). Construction is a complex and multifaceted industry in which innovation seems to present many faces, most of which are still little explored.

We then move to present the current situation of rapid change in the construction sector of the economy in Chapter 3. Its significance to the wider economy and the importance of its impact on and contributions to organizations and to society as a whole is being rethought. Issues of low productivity have brought about a closer examination of the industry. This has led analysts to redraw the boundaries of construction on a broader basis. This reconceptualization addresses the need to include aspects of construction, such as engineering services and suppliers rather than just contracting or construction-site related organizations. These give a more appropriate basis for comparing construction productivity with other sectors such as automobiles, which have always included design as part of the sector whenever measurements were taken. This has renewed interest in the question of the key players, their roles and their capacity for change and for acting as agents to promote innovation. André Manseau presents the key drivers of change in the industry and the emerging importance of service aspects of construction, such as long-term maintenance contracts.

As one response, Frances Anderson begins with the issue of how to define construction activities and thus define who is involved in construction and how we measure innovation (See also Winch 2003). Her chapter discusses attempts to statistically survey the construction industry using the case of Canada and a 'construction cluster' approach based on the OECD model (above). Of particular interest is the question of the relationship between different types of transactions or interactions and the diffusion of innovations across the firms involved in the construction sector. The findings of the survey are presented to show the determinants of innovation in Canadian construction and related industries.

Recent models of complex, connective processes are considered in detail by Graham Winch in Chapter 5. New approaches to the management of innovation and new understandings of the innovativeness of entire industries are considered. Graham Winch begins with industry life-cycle models before moving on to discuss the relevance of new approaches for construction. UK approaches to Complex Products Systems (CoPS) are reviewed, followed by analyses of the role of technical organizations such as standards bodies and trade associations, inter-firm collaborations, gaming models of innovation in different industries, the geographical and urban bases of social capital, and Japanese models of technology fusion. In all of these, the role of brokers and champions of innovation appears crucial but still is left unclear, which sets an agenda for further research on the role of third parties in different industries and in different innovation contexts.

In Chapter 6, Roger Miller and Serghei Floricel sketch out strategies they have found for coping with the risks and high stakes from their analysis of large engineering projects. The seeds of success are planted early on as successful sponsors attempt to influence the drivers of risk even before the project has taken final shape.

André Manseau and Elisabeth Campagnac consider the role governments and state organizations might play in support of construction innovation. As clients, building owners and as regulators, governments clearly have a major impact on and stake in construction. State approaches to the private-sector construction industry have been strongly influenced by economic and political ideologies. Drawing in part from international comparative research (Manseau and Seaden 2001), many different approaches to government intervention can be found at the present time. However, the same economic and social trends and environmental challenges can be found across these countries. National governments share a common set of directions in their policy responses. A focus on building social capital suggests a set of roles for governments and public policy instruments to accomplish national goals.

Based on a bibliometric analysis of job advertisements, Andre Dorée and Frens Pries argue that a long-standing mind-set or management paradigm continues to dominate in large construction firms. This chapter surveys resistance to organizational change and to new alignments and approaches within construction and land development companies.

Chapter 9 turns to the role of skilled labour and organizational culture in innovation. Health and safety have long been drivers of innovation. However, the competencies and expectations of workers in construction are changing dramatically away from collective, often manual, work to isolated and often service type work or machine operation done by individuals working alone. The value of the social capital of skilled trades organizations has been negatively affected by innovation and change in construction, resulting in low morale, high turnover, low recruitment into trades despite historically high rates of remuneration and accidents. This poses one of the greatest challenges to construction and related industries.

The Conclusion reports major trends, key lessons that have been learned and main drivers and barriers to innovation. This chapter is the result of a roundtable discussion of the contributors and invited experts, including George Seaden, Sherif Barakat, Keith Hampson and Roger Miller. Fundamental changes in the way of doing business and the relationship to clients are driving changes in the culture and governance of construction firms. The implications of these changes and future directions are assessed to give a sense of the agenda for researchers, industry practitioners and policy makers.

This book would not have been possible without the support of many people, especially at the National Research Council of Canada. In part, this book grew out of research on Cultures of the Construction Site funded by the Social Sciences and Humanities Research Council of Canada. The editors acknowledge the contribution of graduate students at Carleton University. This book would not have happened without the early help of Jane Hampson, research manager at the Innovation Research Unit who wrote many letters, tracked down contributors and managed the budgets of several research projects during the life of this project. Thanks are also due to Ellen O'Halloran who proofread the manuscript and transcribed the Roundtable. Finally, Cathryn Kallwitz conducted research for Rob Shields, made many final corrections to the contributions, and constructed the Index and camera ready copy.

References

Manseau, A. and Seaden, G. editors (2001), Innovation in Construction – An International Review of Public Policies, London: Spon Press.

Winch, G. (2003). 'Models of manufacturing and the construction process: the genesis of re-engineering construction' *Building Research and Information* 31(2), 107-118.

Chapter 1

A Survey of the Construction Innovation Literature

Rob Shields

1.1 Introduction

This chapter offers a critical overview of research on innovation and its application to construction research. It does not seek to be encyclopaedic both offers a sketch of the main tendencies and major contributors to the English-language literature over the last 30 years. Many of the points sketched here are taken up in more detail in later chapters, for which this Introduction provides a general background.

A famous distinction contrasts invention to the generation of new ideas, with innovation, the application of new ideas (Shumpeter, 1976). Van de Ven comments that innovation is a matter of managing 'new ideas into good currency' (Van de Ven, 1986 cited in Winch, 1998: 269). But in the literature on construction, a certain mystery surrounds innovation. On the one hand, there have been significant changes in the way in which the built environment is produced, such as the reliance on cranes and other mechanized approaches (Shapira, 1996). On the other hand, construction retains the image of a craft industry which appears to defy attempts at rationalization and standardization. Constructed facilities consist of complex and interdependent systems that are often required to function without maintenance over long periods of time, while being subject to unpredictable or harsh environmental events. 'Introducing change into this complex multi-system context can create a ripple effect of secondary and tertiary impacts, which can be difficult to anticipate using current construction management theory and techniques' (Slaughter, 2000:2-3).

As such, construction and its related industries share characteristics of what have been called 'complex product systems' (Miller, Hobday, Leroux-Demers, & Olleros, 1995). Winch summarizes such industries as: (a) having many interconnected and customized elements which must be fitted together according to a specific and sometimes hierarchical scheme; (b) non-linear and emergent (i.e. unpredictable) qualities, and (c) a high degree of user involvement in innovation (Winch, 1998:269; see also Chapter 5). In the case of the built environment, to this one we might add the involvement of public regulatory, community and interest groups, clients and financial institutions.

Even if change is possible, consider an engineer who proposes a new approach to a specific component or even an overall project. First, the problem is likely to have been recognized intuitively based on the innovator's own preoccupations. Many industries 'are based on understandings which began in the making of objects' (Groák, 1992: 179; see). The solution is likely to be narrowly based on the innovator's own field of competence or knowledge of a particular aspect of building, rather than monitoring and evaluation of the whole building over its lifespan.

In past centuries this did not appear to matter as sufficient knowledge of building behaviour was gained over long periods in which methods remained fairly stable and piecemeal improvements could be tested. This affable method of historical patience – and its associated reliance on precedent – is no longer adequate for today's social, economic and technical change and the consequent rapid onslaught of new demands upon both new and existing buildings (Groák, 1992: 53).

Today, building regulations govern the layout and performance of its elements. Safety and labour codes interlock with union certification to establish who performs tasks and how work under this division of labour is conducted. Novel equipment and materials may be embargoed due to customs regulations, and time to test and develop effective practices may not be available. Cost, risk, uncertainty and control over only limited aspects of the way that construction work is actually done – and lack of knowledge, skills and experience in dealing with other areas of construction and engineering – makes change difficult. Thus, as Winch wryly notes, at times it is possible to have both too much and too little innovation: there may be plenty of new ideas, new products or processes, but they may not achieve 'good currency' and be disseminated or taken up in more that one project (Winch & Carr, 2000)

Such uncertainties and risks are deliberately shifted back and forth between actors in the construction process – suppliers or manufacturers, installers and owners. Often risk is resolved through networks of trust, rather than merely economic transactions (Shields, 2003). Syndicating risk, for example through insurance, is relatively rare or borne by the State which may provide workers' accident and health care through national workplace health and safety legislation and injury-compensation schemes. Trust and confidence is built through ongoing interactions which allow actors within the network of the construction industry to anticipate the performance of other actors under a wide range of conditions. One aspect which is less studied, is the informal system of governance which overlays formal institutions, regulations and contracts in construction. Innovation unfolds within this dual structure of governance and within a dynamic of both cooperation and competition between actors (see also Miller et al., 1995).

Innovation certainly does occur in construction and related industries but there is a negative stereotype that the rate of innovation lags behind other sectors. The building industry has adapted and innovated constantly but formal investment in research and development is low. Others argue that its rate of productivity growth is low or even loosing ground in comparison to other industries (Seaden, 1996). But because regulatory frameworks impinge on the forms projects may take (fire safety, zoning, occupancy codes), and the processes by which they are built (labour, health regulations), innovation takes place with regard to more than competitive advantage and reputation alone. Innovators also have their eye on the operational advantage of ease of implementation and reliability with respect to liabilities and warranties (see Lampel, Miller, & Floricel, 1996:368).

However, research itself has perpetuated this stereotype by defining the construction industry narrowly as primarily on-site assemblers and the immediate suppliers. Many manufacturers produce products for a variety of sectors and end up being defined as non-construction firms in national statistics. By contrast, construction is grouped with broadly defined industries such as automotive or aerospace. Statistics are collected from not only assemblers and immediate suppliers but also sales, leasing and service sectors of these industries (see Ch. 2).

As early as the 1960s Marion Bowley pointed out the distinctive character of innovations in British construction and civil engineering. In an overview of changes in structural practices since the mid 1800s, she noted that the construction industry responded above all to the disappearance or unavailability of familiar materials and demands for entirely new types of building and structures. Considerations of cost or the availability of new materials and approaches did not figure.

When the sort of building currently erected becomes technically inadequate for the purpose for which it is used, the resulting new problem tends to stimulate innovations and attempts to develop them. It appears that the important point is the posing of a defined problem, for it is also obvious that radical decreases in the availability of familiar materials or other resources equally stimulate the development of major innovations in structures.

On the other hand the possibilities of either improving the technical efficiency of structures, or reducing their costs, have not tended to be powerful influences encouraging innovation in this country... Similarly, actual increased availability of resources appears to have relatively little influence. (Bowley, 1966:169-170).

Furthermore, she observed that the building industry 'appears to depend to a considerable extent on other industries for the development of techniques, rather than itself (Bowley, 1966: 170). Not only is construction a recipient of techniques but the innovations diffuse unequally in time and space, with time lags and dependencies on foreign technical innovations.

Even when the main principles had been worked out and demonstrated in other countries, there was a singular lack of interest and an unwillingness to master the new techniques of design. It was noticed that engineers were particularly resistant to the use of reinforced concrete, which involved a greater breach than did the steel skeleton with the existing techniques of design of composite buildings of masonry strengthened with iron or steel, or other iron and steel structures.

> ...the stimulus of new problems is required. These new techniques did not provide any solution to new problems in this country; new buildings capable of serving new needs were not urgently required by prospective building owners (Bowley, 1966:172).

New materials for structural purposes do not necessarily stimulate innovation, at least in the UK. Innovation in materials is a necessary but not sufficient cause of innovation. The time lags in the take up of new approaches and materials points to the important role of a requirement for new types of building in which new materials provide a solution. Cost efficiency does not appear to be a factor until after the Second World War. Changes in taste and aesthetics also have little impact on the course of construction. Moreover, 'aesthetic innovations have, with few exceptions, tended to run counter to, or to be irrelevant to, innovations in structures when not actually incompatible with them' (Bowley, 1966:170).

Even in this historical research from the 1960s, the key elements of many innovation studies are present including:

- Barriers to innovation.
- Diffusion of innovations in time and space.
- Clients or owners as important drivers of change.
- Divisions between designers and builders are noted as a structural barrier to innovation.
- Aesthetic changes in architectural tastes are discounted while,
- Incremental solutions to the need for cost-savings or convenience appear ad hoc and peripheral to major developments in the industry.

These themes recur in studies of innovation in construction over the last 30+ years. These main themes will be used to help organize the rest of this chapter, with references to further detail which can be found in later chapters.

Diffusion of innovations in time and space

In the early 1990s attention shifted to innovation as a basis of productivity improvements in the economy generally. Organizations such as the OECD provide definitions that allowed measures of innovation to be coordinated across countries (OECD/Eurostat, 1997). On the positive side, this has facilitated international comparative research on innovation in construction (eg. Manseau & Seaden, 2001); however even though researchers quickly realized that there were problems in importing a definition of innovation based on experience of manufacturing industries the coordination of research required adherence to a rigid definition of innovation focused on the commercialization of new products and processes by firms rather than clusters of firms working together in project alliances or even procurement innovations by clients. In the case of construction innovation, national panels in a number of countries were able to identify construction as a key to achieving wider social goals such as energy efficiency and reductions in unemployment rates (cf. Latham, 1994; NIST, 1994; Seaden, 1996).

Freeman (Freeman, 1989) is often cited on the distinction between invention and innovation. Invention is 'a detailed design or model of a process or product that can clearly be distinguished as novel compared to existing arts' (Slaughter, 1998: 226). Innovation is a matter of application and operationalization; it is defined as a non-trivial improvement in a product, process, or system that is actually used and which is novel to those developing or using it (Marquis, 1999; Slaughter, 2000). Furthermore, innovations are changes which persist and are significant enough that they are documented. In this manner, and perhaps a bit unfairly, we differentiate innovation from, for example, tacit 'experience' and one-off 'problem-solving' (Shields & West, 2003) which has been held up as an alternative to formal R&D (Groák, 1992):

Construction projects involve considerable problem-solving as the general repertoire of technologies and techniques is adapted and applied to meet the specific client's needs in interaction with the constraints of the site. For problem-solving to become innovation, the solutions reached...must be learned, codified, and applied to future projects - knowledge that remains tacit is difficult to manage into good currency (Winch, 1998: 273).

Others have worried that there is an overemphasis on the distinction between invention, as the genesis of ideas, and innovation, understood as their exploitation.

It suggests that the genesis of ideas (discovery) is somehow exempt from the influence of the social networks that are so important for exploitation ... whether or not something counts as an idea, and whether or not that idea counts as new, necessarily depends on the social networks involved. Somehow, somewhere, someone has to be convinced that it is a 'new idea'... (Woolgar, 1998).

1.2 The Scope of Innovations

Slaughter offers a range of five types of innovation. These expand upon the difference between major developments and the ad hoc micro-innovations that are a part of fitting every project to the new and unique conditions of a particular site or the needs of a different client (Slaughter, 1998):

Incremental innovations most often appear within organizations that have a knowledge base on which to develop improvements. These are small improvements with minimal impacts on other systems (Marquis, 1999). An example might be the improvement of safety harnesses. Such off-site innovations are driven by cost-saving and convenience requirements but can be spurred on by incremental changes in regulations. On-site, incremental innovations are part of the folklore of construction - they circulate as rules of thumb and apocryphal stories of the origins of construction practices in response to unique conditions or client requirements.

Modular or Product innovations are improvements within a specific system which requires no changes in other components or systems (Henderson & Clark, 1990). These are amongst the most common innovations within construction, as they can be implemented within a single organization without requiring a renegotiation of the construction process. These are often product innovations that are the result of R&D activities by manufacturers or the introduction of new products by large suppliers. Indeed, the producers of materials and components carry out the majority of construction-related research. Examples might include improvements in fire-rating or the introduction of improved building components on the basis of new technology internal to the component itself. But Gann and others argue that the construction industry tends to focus too much on product enhancement rather than process innovations because of the structural challenges of a complex industry and complex, site-specific end-products (Gann, Matthews, Patel, & Simmonds, 1992; Winch, 1998:269).

Process Innovations include new ways of assembling buildings and the reorganization of practices, which may be required by new components. At their grandest, so-called *architectural or configurational innovations* require changes in other activities on which they depend or are linked to (Hughes, 1983; Teece, 1988) . Often this is a result of one or more phases of the construction process being eliminated or its timing changed. There are greater commercial advantages but the barriers to success are higher because the implications for other building systems have to be understood and new linkages in the project schedule and between tasks must be worked out. There is some evidence that these innovations originate in the construction field itself (Slaughter, 1993) despite the complaint of Gann et al above that the field focuses instead on product innovation.

Process innovation points to the manner in which all engineering and construction projects are complex feats of organizing people and objects. They have to be fitted into the time flow of a project, the space of a site and the historical time of a built object or environment. There is an ongoing process of practical 'doing' and learning. This includes persuasion, negotiation with project partners and the management of persistent contradictions, uncertainties and risks (Suchman, 2000: 325).

An example of the outcome of such learning processes and negotiated arrangements over time can be found in the shift to mobile cranes (truck-mounted and self-propelled). Because of a significant number of relatively homogeneous new construction projects of short construction duration, on spacious sites and supportive terrain, a tendency to use mobile rather than tower cranes can be found in the South Western United States. Non-project-specific factors such as planning culture, operating style, employment patterns, and market organization have also shaped what Shapira and Glascock call a 'mobile crane culture' (Shapira, 1996) which avoids the use of tower cranes which are traditionally identified with high-rise, congested new construction in urban sites.

System innovations are sets of interdependent new products and practices. These multiple innovations are linked and complementary (Cainaraca, Colombo, & Mariotti, 1989). Some also refer to 'architectural innovations' - those requiring modification in other components or systems (Henderson & Clark, 1990). These are qualitatively more extensive than more limited changes in the configuration of components. Characteristically, these innovations involve a process of integration and complementarity based on new technologies. As in the case of widespread computerization and automation, they reconfigure existing interests by competing with existing components, systems and trades. Centres of responsibility and control are often consolidated. An example would be the case of the switch to interior 'drywall' rather than wet plastering which reduces plastering skills and radically changes the system of transportation and wholesaling of interior finishes.

Radical innovation is a completely new concept of approach which renders previous approaches and interdependent components or systems obsolete (Nelson & Winter, 1977). Radical innovations such as the introduction of steel or structural use of concrete are reflexively tied to the emergence of new industries, new construction systems and construction specialties. Slaughter notes that these appear from outside of construction (Slaughter, 1998:228). Yet a degree of ambiguity remains in the literature, because the same innovation could appear as a small, modular improvement to a specialized construction firm and a radical innovation to the majority of general contractors.

Prefabrication is one construction process that allows such combinations of systems and of different types of innovation to be implemented. System innovations involve a change in not only products and assembly practices, but in approach and conception. Standardization is one such innovation which is pervasive both onsite and across the breadth of construction - related industries and regulations. However, the degree of standardization has waxed and waned. At the scale of the entire building it has been successful only in industrial applications (such as steel arch buildings on the model of the 'Qwonset Hut'), but at the scale of components and standard assembly procedures it has been widely successful, largely depending on the degree of hegemony enjoyed by dominant suppliers.

The literature on construction innovation is full of typologies. But the exercise of classifying innovations is a crude, static tool that provides little insight into either specific innovations or the dynamics of construction systems. Such classifications are based on hindsight, always post hoc. They are didactic but of little use to those confronting new innovations. A change implemented as a 'modular innovation' may lead to changes in other aspects of construction - only in this manner would it show itself to be a radical innovation.

The problem with classifications is that they overlap and may combine sets of distinct innovations. Worse, they depend on one's perspective - the position with respect to which a change is examined. It thus becomes very difficult to evaluate innovations. Slaughter admits that 'while an innovation may be perceived to be an incremental innovation to the project members on the site, it may have required extensive off-site development and preparation and indeed may be perceived as a significant change to the state-of-knowledge and the links to other components and systems to other members of the value-added chain' (Slaughter, 1998:8; see also Afuah & Bahram, 1995). The perspectival character of innovation leads to the conclusion that it is necessary to evaluate any given change from every possible position in the network or social-technical system from suppliers to users, including related building components. Furthermore, any given change is potentially all types of innovation at once, depending on whose perspective one adopts. In practice, perspectivalism sets the bar for assessments too high to be useful.

Research on innovation must confront the paradox that neither categories nor the notion of 'innovation' itself provides any predictive leverage. Changes appear as innovations only in hindsight. This post hoc paradox means that the scope and significance of the innovation is also only clear in hindsight. By definition there are no bad innovations, and innovations are never failures either in terms of market success, dissemination or functionality. Innovation is in fact change that has been deemed successful (and usually profitable) for its protagonists. The story of the unsuccessful innovation is of interest only to the anthropologist. In general, the literature recommending various innovation-related policies and management strategies appears to be little more than a compendium of wishful thinking in which the experience of those negatively affected by innovations is suppressed.

The weakness of this post hoc approach is that it describes the past in categories which may not easily fit the circumstances of an unfolding present. An incremental innovation may unexpectedly be complemented by another, apparently incremental innovation to suddenly offer the opportunity for a system innovation. The outcome can only be known in hindsight. Worse, it offers little strategic insight in adopting one or another innovation.

Perspectivalism and post hoc paradox are at the root of the difficulty of identifying and assessing innovations. As a result, identification of innovation in construction relies on anecdote. This gives the literature a folklore-ish quality, more a set of moral tales and images than rigorous social science. How are we to advance in such a situation? On the one hand, some have attempted to measure what is quantifiable, forging ahead despite the vagueness of underlying definitions and with an implied acceptance of the authority of 'native informants' - construction supervisors, foremen, owners and managers of engineering contracting companies or suppliers. In survey methodologies, there is little opportunity to question this authority; the categories constructed by outside researchers or to explore the respondents' interpretation of questions with which they are faced

1.3 Innovation Processes

There is an extensive literature on construction innovation dispersed across a number of subfields. These include engineering management, architectural history, macro-economics, and research policy-making. Business schools offer courses in 'innovation' but few read the literature specific to construction, archived in specialist engineering publications.

To better understand the details of innovation, empirical research has mapped the process and key management tools for ensuring the successful implementation of innovations (Winch & Carr, 2000). In the 1980s and early 1990s, much of the research focused on innovation within firms. The role of 'innovation champions', the introduction of new technologies and solutions to professional problems were emphasized as sources and forms of innovation. By contrast, in the late 1990s, attention shifted to the competitive environment in which firms operate, and intra-firm linkages and collaborations which were necessary to successfully integrate innovations into the flow of complex projects (Shove, Packwood, & Shields, 1997). The importance of suppliers and the traditional construction industries' apparent function as an assembler of innovations developed elsewhere shifted the focus away from engineering firms as the predominant innovators. The role of clients as participants in a distributed innovation process network rather than as recipients of innovative construction, as users of an end-product, challenged unidirectional visions of the flow of innovation.

Even beyond the innovation process, the problem with linear models of innovation is that they presume that innovation itself takes place in a linear chain, 'downstream' from invention and research. Both analysts and policy makers seem to imagine that 'there is a (rational) path of connection between university or lab and contractor or designer which has been blocked' (Woolgar, 1998:442) and merely has to be unblocked.

Many countries have national construction research labs that have played a vital role in supporting innovation centred around products and materials. These facilities, such as the Building Research Establishment in the UK or the Institute for Research in Construction in Canada have provided the infrastructure to test new components and to develop products in relation to health, safety and performance regulations. However they now almost entirely work responsively with industry - the linear approach has been more typical of industrial policy models than construction research and project management. As Bowley observed, demand - side factors have most often dominated discussions of construction: clients bring projects and problems to which innovation is a response (Bowley, 1966).

1.4 Clients or Owners as Drivers of Change

'Clients have come to the fore in the construction management debate and practice over recent years' (Gibb & Isack, 2001: 46). The importance of demanding clients is often noted in the literature on change in institutional and commercial construction and engineering. But at the same time, in residential construction, the opposite view prevails. Nam and Tatum go as far as to argue that 'the notion that owner's demands or problems dominate the innovation process in construction is largely a myth' (Nam & Tatum, 1992:507). For example, new home buyers in Britain in the 1980s were characterized as,

> a fragmented and passive group. They generally lack information about energy saving and there is a self-reinforcing consensus about what a conventional house should look like and how it will perform. With the exception of housing associations, few organizations have attempted to survey customer views on energy efficiency. Developers note only that houses are sold in ignorance of energy performance. Most people purchasing their own homes are unlikely to attach great importance to energy efficiency because expenditure on energy is usually very low: the average household spends about 4.5% of all its expenditure on energy and only around half of this is on heating. These conditions tend to result in an imbalance of information on user opinions, a real and perceived lack of user demand for more energy efficient housing and a lack of consequent builder demand on suppliers for energy efficient products with which to construct new housing. In general the sector is risk-averse and tends towards a conservative approach with respect to new technologies and marketing strategies (Barlow & Bhatti, 1997).

Conservatism in the development and construction process increases the significance of upstream manufacturers and suppliers and of downstream users in innovation (Gann, Wang, & Hawkins, 1998:287). The situation changes as homeowners expect greater comfort and expend more of their income on heating and cooling. Complacent purchasers and users, however, lead the construction industry to maintain a preoccupation with aesthetic values and traditions rather than innovation. But the above quotation points to the importance of information and learning amongst not only engineering and construction industry actors but also amongst clients. From this perspective the situation is more nuanced. In the case of large infrastructural projects where client/owners have an internal technical capacity and thus the ability to both absorb technical information and intervene in projects at every stage, those with an engineering design capacity tended to fetter their contractors, resulting in less innovative projects. Those with only an internal project management capability were generally more innovative (Lampel et al., 1996:366). Clients who fully delegate decision-making to contractors tend, as one might expect, to receive projects in which the prime contractor has innovated with respect to initial cost rather than long-term performance. As part of project management capacities, construction management capability enhances clients' ability to evaluate long term versus short-term benefits and to foresee problems involved in implementing an innovation.

Detailed research on the case of large engineering construction projects showed that innovation occurs as the result of partnerships and joint problem solving by owners, prime contractors, consultants and equipment suppliers.

It is rare for a construction professional, without obtaining a contract, to first conceive an innovative idea and then try to find a buyer for his plan. As compared with the user's passive role in the manufacturing industry, the importance of the owner's initiation, or the customer's order, is believed to be one of the peculiarities of the innovation process in construction.

> When construction innovations are examined, a very few of them fit, in an unambiguous manner, this widely believed model of the innovation process. This is to say that it is extremely difficult to describe the majority of the cases in terms of a linear sequence with the owner's demands as a clearly defined starting point of the innovation process. In many cases, indeed, owners' demands are responsive rather than initiative, it seems that one of the factors that differentiates innovative projects from non-innovative ones is the overturning of the conventional belief that owners' demands should always come first. (Nam and Tatum, 1992:509).

In some cases, strategic partnerships were extended from very large clients to the level of sub-trades in order to achieve faster construction times or to meet entirely new building needs (Shields & West, 2003). Information asymmetries between participants were a major barrier to the successful introduction of new products (Lampel et al., 1996). Build-own-operate partnerships and profit-sharing agreements represent strategies for aligning the interests of contractors with clients. The result is argued to be a bias toward quality, long-term approaches to maintenance and building performance as well as allowing clients to focus on their own major business strengths. Agents are 'more likely to favour solutions that contribute to long-term project performance' (Lampel et al., 1996:366). By the same means, contractors shift the risks of innovating toward clients (Freeman, 1989).

1.5 Construction as a Divided and Complex Systems Industry

Viewed as a complex systems industry, construction consists of an innovation superstructure of clients, regulators and professional institutions, and an innovation infrastructure of trade contractors, specialist consultants and component suppliers (see Ch. 5 and 6). These are mediated by a set of systems integrators who are what is most often thought of as the 'construction industry proper', architects, engineers and prime contractors. This group forms a type of interface between the two by drawing together knowledge and codes, on the one hand and materiel and labour on the other. Winch argues that construction has not one but two systems integrators, one at the design stage (designers) and one at the construction stage (contractors) which is rarely given full authority or takes full responsibility, the result is 'divided professional groups (architects versus engineers), and a set of activist *innovation brokers*' (Winch, 1998:270).

An important distinction of complex systems industries such as construction is that although decisions to adopt a new idea or approach are made in firms or by teams of firms, many innovations must be implemented in projects, not within the firm itself. Projects are collaborative engagements with other actors, and hence the firm's position in this industry network over time and its power to act as an innovation broker influences the chances that an innovation will be adopted more widely.

Divisions between key actors including engineers, contractors, fabricators and suppliers are a common complaint. In much of the literature, designers are removed from the innovation process, which appears as an innovation 'push' from component suppliers through a reluctant group of assemblers or systems integrators (the construction industry, narrowly defined) responding to the 'pull' of the demands of savvy clients. However, Nam and Tatum identify technically competent and progressive designers as key *champions* of architectural and system innovations which integrate components or require construction actors such as sets of suppliers, general and specialty contractors and trades unions to innovate together (see also Pries & Janszen, 1995). Furthermore, public interest groups and regulatory agencies must acquiesce to new approaches (Nam & Tatum, 1997:262-3). Champions are indispensable sponsors, protectors and promoters of innovative processes and products who are able to negotiate and balance the opportunities of new approaches with the risks of departing from the tried and tested.

Nonetheless, the split between architects and engineers as the designers and contractors as assemblers means that championing and brokering must be shared:

> While principal architect/engineers typically display competence in the regulatory framework and client requirements, they often do not have the skills to integrate the subsystems... [In the UK], they are not well complemented by the principal contractor whose integration capabilities are

typically restricted to the managerial rather than technical level (Winch, 1998: 275).

Slaughter finds that builders' innovations arising from on-site problem solving explicitly integrate component innovations into the total building system. Yet these innovations are rarely captured, not only by contractors themselves but by component manufacturers and suppliers (Slaughter, 1993). The problem of technical deficiencies amongst contractors raises quality problems in projects. In addition, however, it impedes the ability of firms to learn by technically comparing and evaluating on-site problem solving and evaluating opportunities to implement new products or approaches (Connaughton, Jarrett, & Shove, 1995; Shove & Shields, 1996). Thus experience remains tacit at the level of site supervisors and project managers (Shields & West, 2003) and does not build the reputation of firms for excellence in particular construction niches.

Despite arguments to the contrary, professionals and their institutions are conservative by nature. They may slow innovation in their quest for control over the building process and contractors or engineers. Resistance may come through reluctance to support changes to regulation or lack of interest deriving from an innovation not lending privilege or jurisdictional advantage to a particular group. Veshosky found that project managers were often unaware of their firms' policies or programs intended to assist them in obtaining innovation information, or did not use the company libraries that might be expected to be sources of information. Although changing with the impact of the World Wide Web, trade magazines and conversations with colleagues about their experiences with innovations were the primary source of information and learning opportunity (Veshosky, 1998). Nonetheless others have argued that project managers tend to see their projects as unique and thus are less likely to consider the lessons learned from other projects (Kahneman & Lovallo, 1993). These lessons must be contributed by experienced consultants (Lampel et al., 1996:369; see also Katz & Allen, 1982).

Again, it is the client, offering the inducement of a commission, who is in a position to push for change by directing particular systems integrators to provide a solution not otherwise available. While individuals may be champions and particular players such as main contractors or engineering firms can be said to be brokers of innovation, clients again find a place as *mediators* of innovation processes. Many have argued that there is a lack of incentive to innovate in construction: the literature is dominated by discussions of barriers, risks of technical failure and liability. Gain-sharing approaches that break the mould of zero-sum scenarios for change are required. When driven by clients, case studies show that innovation is a result of gains the client expects to make as a result. Successful and enduring innovations require that rewards be shared between the players involved in the construction process, including clients.

1.6 Structural Barriers to Innovation

Building codes and construction regulations, both within jurisdictions and in major export markets (such as, for example, the United States is for Canada or Mexico), are often cited as constraining or driving innovation. For Porter, safety and environmental standards can pressure firms to innovate and upgrade technologies (Porter, 1990). A shift to 'performance-based' building regulations was a feature of the last 15 years. However, Gann and others have found that in order to stimulate systemic innovation, performance – based codes must take the form of standards which are flexible enough to allow firms the freedom and institutional frameworks in which to innovate in response to market opportunities (Gann et al., 1998). In effect, initiatives to enhance the competitiveness of firms in various construction and engineering sectors are often contradicted by other regulatory policies. Many have pointed out that regulations need to accommodate technical change at different levels in the production process, including new product development and systems integration. In broader terms, however, the lack of coordination of policy frameworks reflects the contradictory social and economic objectives which the construction sector must satisfy. Research has pointed to the importance of external factors such as the economic environment. When positive, construction innovation appears to flourish (Pries & Janszen, 1995).

Table 1.1 Impediments to innovation (after Bernstein & Lemer, 1996: 87)

General impediments
- Tort liability, threat of litigation and high cost of insurance

Structural characteristics
- No single government agency in total charge of construction
- Large numbers of small firms, operating in limited markets
- Multitude of regulatory codes and standards
- Long service lives of facilities and their components
- Cyclical downturns in design and construction markets

Cultural factors
- Procurement policies, particularly in the public sector, that emphasize lowest initial cost rather than best performance.
- Divided and often adversarial views of labour and crafts participants
- Reluctance of private firms to invest in experimentation, research and development for longer term profit
- Strong reliance on past experience, attitude of 'If it ain't broke, don't fix it!'
- Pervasive public attitude regarding construction, 'Not in my backyard!'

These impediments to innovation are difficult to change because they relate to a wide range of social processes that extend beyond even the broadest definitions of construction. They are thus structural in the sense that they appear as a socio-economic framework within which construction as a general social activity must operate. Shifts in tort liability, for example depend on legal reforms that would require the creation of broad social coalitions to accomplish.

1.7 Summary

While providing an overview of the recent literature on innovation and construction this chapter has sought to give a critical sense of the extent to which research on construction innovation has been preoccupied with many of the same questions for over 30 years. These questions include regulations, risk, client preferences, and the complexity of both the industry and of buildings. New definitions of innovation and understandings of its importance in the generation of wealth periodically sweep construction research, often from other fields such as manufacturing, high technology, or from industrial and macro-economic policy research. These provide a series of new insights but often founder on issues and problems identified in earlier research. These features, such as the importance of project-based work, or the collaboration of many economic actors, large and small, differentiate construction and related project-based engineering sectors from popular stereotypes of mass-production manufacturing or of service industries.

Problems with research on construction innovation arise because construction is defined too narrowly as on-site activity, rather than an entire network of activities. A second problem arises because of the tendency of economic theory and of policy making to focus on firms rather than project teams. In the research, there is disagreement over the classification of innovations. These tend to be post hoc definitions that are not strategically useful as guidelines to the implementation of future innovations. These classifications are also highly perspectival in nature – innovation is defined as things: products or processes new to a specific player in construction rather than an ongoing process that spans specific products or the adoption of a specific business process. Drivers and barriers to specific innovations thus characterize the literature. However, more recently a shift towards industry-wide and entire supply-chain processes can be found. The emergence of a conception of the industry as a complex system within which different groups and actors struggle for advantage offers a more socially-nuanced portrait of innovation which must always be in the interest of some and against the vested interests of others.

References

Afuah, A.N., & Bahram, N. (1995). The Hypercube of Innovaiton. *Research Policy, 24*(1), 51-76.
Barlow, J., & Bhatti, M. (1997). Environmental performance as a competitive strategy? British speculative house buildings in the 1990s. *Planning Practice and Research, 12*(1), 33-44.
Bernstein, Harvey M., & Lemer, Andrew C. (1996). The Innovation Puzzle, *Solving the Innovation Puzzle: Challenges Facing the U.S. Design and Construction Industry* (pp. 87-107). New York, N.Y.: ASCE Press. American Society of Civil Engineers.
Bowley, M. (1966). The Development of Modern Building Structures and Industrial Change - Conclusions, *The British Building Industry, Four Studies in Response and Resistance to Change* (pp. 169-180). Cambridge: Cambridge University Press.
Cainaraca, G.C., Colombo, M.G., & Mariotti, S. (1989). An Evolutionary Pattern of Innovation Diffusion: The Case of flexible automation. *Research Policy, 18*(1), 59-86.
Connaughton, John, Jarrett, Neil, & Shove, Elizabeth. (1995). *Innovation in the Cladding Industry*. London: Department of the Environment.
Freeman, Chris. (1989). *The Economics of Industrial Innovation*. Cambridge, MA: MIT Press.
Gann, David, Matthews, M., Patel, P., & Simmonds, P. (1992). *Construction R&D: Analysis of private and public sector funding of research and development in the UK construction sector*. London: Department of the Environment.
Gann, David, Wang, Yusi, & Hawkins, Richard. (1998). Do regulations encourage innovation? - the case of energy efficiency in housing. *Building Research & Information, 26*(5), 280-296.
Gibb, Alistair G.F., & Isack, Frank. (2001). Client drivers for construction projects: implications for standardization. *Engineering, Construction and Architectural Management, 8*(1), 45-58.
Groák, Steven. (1992). *The Idea of Building*. London: Spon.
Henderson, R.M., & Clark, K.B. (1990). Architectural Innovation: The Reconfiguration of existing produce technologies and the failure of established firms. *Administrative Science Quarterly, 35*(1), 9-30.
Hughes, T.P. (1983). *Networks of Power: Electrification in Western Society 1880-1930*. Baltimore: Johns Hopkins University Press.
Kahneman, D., & Lovallo, D. (1993). Timid Choices and BoldForecasts: A Cognitive perspective on risk taking. *Management Science, 39*, 17-31.
Katz, D., & Allen, T.J. (1982). Investigating the Not Invented Here (NIH) Syndrome: A Look at the performance, tenure and communication patterns of 50 R&D project groups. *R&D Management, 12*(1), 7-12.
Lampel, J., Miller, R., & Floricel, S. (1996). Information asymmetries and technological innovation in large engineering construction projects. *R&D Management, 26*(4), 357-369.
Latham, M. (1994). *Constructing the Team*. London: Dept. of the Environment, UK.
Manseau, A., & Seaden, G. (2001). *Innovation in Construction: An International Review of Public Policies*. New York: Spon.
Marquis, D. G. (1999). Anatomy of successful innovations. In M. L. Tushman & W. L. Moore (Eds.), *Readings in the Management of Innovation* (pp. 79-87). Boston: Ballinger Publishing.

Miller, R., Hobday, M., Leroux-Demers, T., & Olleros, X. (1995). Innovation in Complex Systems Industries: The Case of flight simulation. *Industrial and Corporate Change, 42*(2), 363-400.

Nam, C.H., & Tatum, C.B. (1992). Strategies for Technology Push: Lessons from Construction Innovations. *Journal of Construction and Engineering Management, 118*(3), 507-524.

Nam, C.H., & Tatum, C.B. (1997). Leaders and champions for construction innovation. *Construction Management and Economics, 15*, 259-270.

Nelson, R.R., & Winter, S.G. (1977). In Search of a Useful Theory of Innovation. *Research Policy, 6*(1), 36-76.

NIST. (1994). *Rationale and preliminary plan for federal research for construction and building* (NISTIR 5536). Washington, D.C.: National Institute of Standards and Technology, Department of Commerce.

OECD/Eurostat. (1997). *Proposed Buidelines for collecting and interpreting Tehcnologicla innovation Data - Oslo Manual*. OECD. Retrieved from the World Wide Web.

Porter, Michael. (1990). *The Competitive Advantage of Nations*. New York: Macmillan.

Pries, Frens, & Janszen, Felix. (1995). Innovation in the construction industry: the dominant role of the environment. *Construction Management and Economics, 13*, 43-51.

Seaden, George. (1996). Economics of innovation in the construction industry. *Journal of Infrastructure Systems ASCE, 2*(3), 103-107.

Shapira, A., and Glascock, J. (1996). Culture of Using Mobile Cranes for Building Construction. *Journal of Construction Engineering and Management, 122*(4), 298-307.

Shields, R. (2003). *The Virtual*. London: Routledge.

Shields, Rob, & West, Kevin. (2003). Innovation in Clean Room Construction: A Case Study of Cooperation between Firms. *Construction Management and Economics, 21*(4), 324-337.

Shove, E., Packwood, N., & Shields, R. (1997). *Factors Affecting Competitiveness in Construction*. London: Dept. of the Environment, UK.

Shove, E., & Shields, R. (1996). *Motivation of Innovation*. London, UK: Dept. of Environment, UK (CRISP).

Shumpeter, J.A. (1976). *Capitalism, Socialism and Democracy* (5th ed.). London: Allen and Unwin.

Slaughter, E. S. (1993). Builders as Sources of Construction Innovation. *Journal of Construction and Engineering Management -ASCE, 119*(3), 532-549.

Slaughter, E. S. (1998). Models of Construction Innovation. *Journal of Construction and Engineering Management - ASCE, 124*(3), 226-231.

Slaughter, E. Sarah. (2000). Implementation of construction innovations. *Building Research & Information, 28*(1), 2-17.

Suchman, Lucy. (2000). Organizing Alignment: A Case of Bridge-building. *Organization, 7*(2), 311-327.

Teece, D.J. (1988). Technological change and the nature of the firm. In G. Dosi & C. Freeman & R. Nelson & G. Silverberg & L. Soete (Eds.), *Technical Change and Economic Theory* (pp. 256-281). London: Frances Pinter.

Van de Ven, A.H. (1986). Central Problems in the Management of Innovation. *Management Science, 32*(5), 570-607.

Veshosky, David. (1998). Managing Innovation Information in Engineering and Construction Firms. *Journal of Management in Engineering*, 58-66.

Winch, G. (1998). Zephyrs of creative destruction: understanding the management of innovation in construction. *Building Research & Information, 26*(4), 268-279.

Winch, G., & Carr, B. (2000). Process Maps and Protocols: Understanding the Shape of the Construction Process. *Construction Management and Economics.*

Woolgar, Steve. (1998). A New Theory of Innovation? *Prometheus, 16*(4), 441-452.

Chapter 2

Construction – A Changing Industry Challenging Current Innovation Models

André Manseau

The construction industry is undergoing an important reconfiguration. Major changes are occurring in this very significant economic sector that has been considered as traditional or mature. New technologies are facilitating off-site production as well as the integration of the design stage to building. New contractual arrangements have been introduced, changing the roles of stakeholders. Global and highly integrated firms are emerging. Former boundaries between manufacturing of building products, design, on-site construction, maintenance and managing existing stock are blurring.

Current statistics show a decreasing industry, but the real picture is rather different. The manufacturing side as well as the services sector in construction is growing. In conjunction with these structural changes, key drivers are significantly specific in this industry. Current economic models of changes and technological development that have been applied in other sectors cannot fully explain reconfiguration and innovations in construction.

2.1 An Industry in Mutation

Construction is a broad industry that encompasses many different sub-sectors as well as a number of public organizations involved in building codes development and regulation of various aspects (safety, fire risk, environment, etc.). The focus on construction activities on building sites has limited the perspective and has blinded economic analysis from major transformations in this sector. We refer to this narrow focus as the 'traditional construction sector'. A larger perspective is needed in order to capture the real picture of the industry. Construction spans from primary industries such as gravel or other aggregates, to service sectors such as geotechnical engineering consultants.

Revisiting current statistics

Limiting the analysis to the traditional sector, construction appears as a declining industry; its share of the economy, in terms of value-added to GDP, has slipped from about seven per cent to five per cent in the last two decades (Figure 2.1).

Construction productivity, in terms of value-added, has stagnated in the last two decades, with the exception of a few countries. The only real improvement in productivity has occurred in the USA (Figure 2.2).

Construction productivity is an indication of its efficiency in producing value-added products or profits. Productivity is the ratio of the industry output or production on its input. National statistics usually provide two measures of productivity: the labour productivity – production per labour input, and the total productivity – production per all inputs (labour, material and capital costs).

We obtain similar conclusions by analysing the trends in labour productivity (Figure 2.3). Although we have productivity measures for different countries, it is difficult to compare productivity levels from one country to another. In particular, the labour cost is not estimated by a common and uniform methodology. For example, some countries estimate labour cost from input data of industry surveys, while others estimate this cost from individual revenue declarations. However, within a country, we should have consistency through the selected time period and therefore, each country trend in and of itself can be considered as robust information.

Productivity levels in the traditional construction sector are still relatively low in comparison with other sectors (Figure 2.4). Again, there are a number of limitations in comparing productivity levels between countries, as mentioned above.

In light of these statistics, some observers might consider construction as a lagging industry. However, construction activities are far than being fully captured by these data. Construction is not limited to on-site production. Recent studies have shown the increasing importance of managing existing building stocks and a general trend toward a rising weight of the tertiary sector of construction such as building operations and maintenance services, and which are not captured by the traditional construction sector. Repair and maintenance represent more than 45 per cent of construction works in many developed countries (Carassus, 2003). Usikyla et al. (2003) estimated that total assets in real estate and construction account for about 70 per cent of the total national assets in Finland. However, it is still difficult to acquire statistical data on these trends.

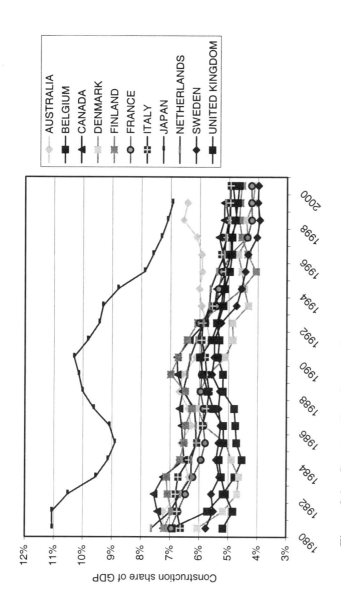

Figure 2.1 Construction share to the economy

Source: OECD, Annual National Accounts – Main Aggregates (2002).

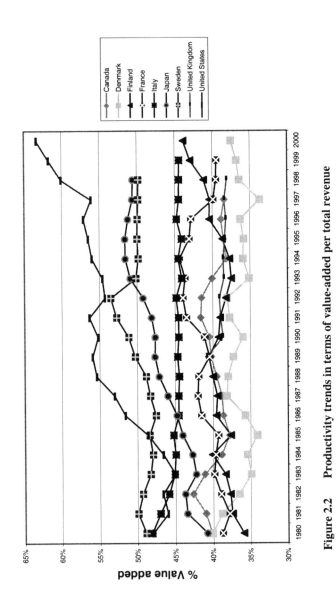

Figure 2.2 Productivity trends in terms of value-added per total revenue

Source: OECD, STAN Industry Structural Analysis (2002).

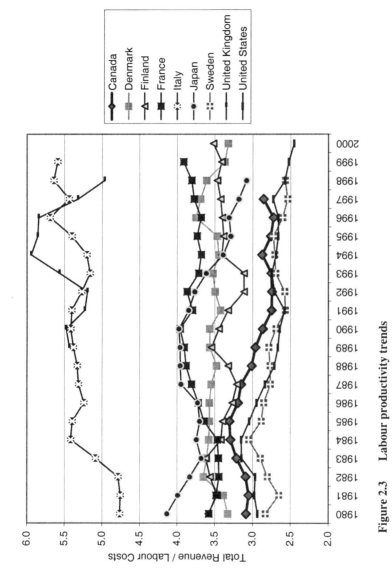

Figure 2.3 Labour productivity trends

Source: OECD, STAN Industry Structural Analysis (2002).

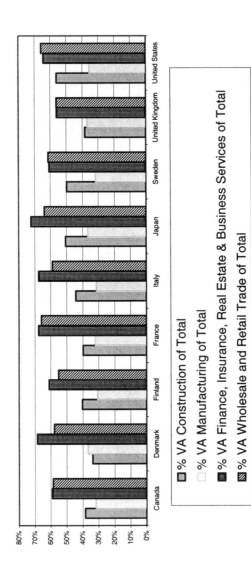

Figure 2.4 Productivity comparison

Source: OECD, STAN Industry Structural Analysis (2002).

Other industry sectors, such as the automotive or aircraft can be measured in a much more comprehensive manner with current statistics. For instance, the International standard industry code (ISIC) enables an easy identification of manufacturers of motor vehicles (code 341), manufacturers of parts for these products (code 343) and even sale, maintenance and repair services of motor vehicles (code 50). In the case of the construction sector, only production on-site can be easily identified (code 45). Assessing the economic importance of construction products and material is very difficult, as they are widely dispersed in a number of manufacturing sectors. It is the same challenge for construction services that are dispersed and not specifically identified within professional services (design, management), within real estate services, and within the 'other services' category (for repair and maintenance).

Redefining construction boundaries

As the understanding of construction has broadened, the number and roles of the different stakeholders/participants involved are becoming increasingly complex. Table 2.1 lists the key players of the industry, as presented by Bernstein and Lemer (1996).

Table 2.1 Key stakeholders in construction (from Bernstein & Lemer, 1996)

Owners	**Designers**
Private companies and individuals	*Engineers*
Government Agencies	*Architects*
Regulated utilities	*Landscape architects*
Stock and bond-holders	*Financial analysts*
Mortgage and other lenders	*Regulatory Specialists*
Insurers	*Others*
Others	
Manufacturers	*Constructors and Operators*
Equipment	*General Contractors*
Building products	*Specialist subcontractors*
Material suppliers	*Craftsmen and labour*
Fabricators	*Maintenance staffs*
Vendors and distributors	*Facilities managers*
	Others
Users	**Others**
End users (commuters, water customers)	*Administrative personnel*
Intermediate users (freight forwarders, food product manufacturers)	*Enforcement officials*
	Others
Neighbours and other impact parties	
Others	

2.2 Clustering Interconnected Players

The importance of encompassing all key players in the construction industry was also stressed by Winch (1998) who identified key characteristics of this complex system:

- Many interconnected and customized elements organized in an hierarchical way;
- Non-linear and continuously emerging properties where small changes to one element of the system can lead to large changes elsewhere in the system;
- A high degree of user involvement in the innovation process.

This evolution has to match a change in the economic analysis and in the scope of the industry. Carassus (1999) has proposed a meso-economic cluster approach for capturing all closely related construction activities. In using this approach we can capture more than the construction industry, which has been the approach of past years, but rather we can acquire a broader range of construction, from manufacturers, designers, developers, builders and to building services.

Mesoeconomics (meso, meaning median in Greek) lies somewhere between microeconomics and macroeconomics (see Table 2.2). Microeconomics is the study of scarcity and how individuals and firms make profit in this market. In addition to this it sets out to explain the equilibrium between sellers, buyers and prices, and how these prices affect the actions of buyers and sellers.

Macroeconomics is at the opposite end of the scale. It sets out to find the equilibrium in national and international markets. To do this it deals with averages such as price average, income average, employment, production, taxes and the effect of government budget spending, and how all these relate to each other. Gross Domestic Product and Gross National Product would be an example of macro indicators.

The mesoeconomic approach does not replace either of these but rather sits in the middle, which the micro approach is too narrow to do and the macro approach is too broad. Not concerned with individuals or the world economy it is more regional. Carassus describes it as putting 'the focus on industry structure in developed economies as well as the political dimensions of economic development and policy formation.

Uusikyla et al. (2003) describe the Finland construction cluster as comprising five sub-fields: Real Estate, Infracluster – including civil engineering, Building, Building products and Building Services. Figure 2.5 presents a systematic approach to the key actors involved in construction who can undertake innovation activities.

Table 2.2 Three levels of economic analysis – micro, meso and macro (as presented by Carassus, 2001)

	Areas and Topics	Elements of Analysis
Microeconomics	Households	Demand Theory
		Cost and price theory
	Firms	Market and Price Theory
		Competition Theory
	Markets	Distribution of Income
Mesoeconomics	Industries	Theory of Economic Structure and Change
	Regions	Regional Economics
		Environmental Economics
		Theory of Groups and Associations
	Groups	Economic theory and politics
Macroeconomics	Macro-aggregates	National Economic accounts
		Economic stability and growth
		Monetary theory
		International trade
	Total Economy	Macroeconomic distribution theory

These actors include:
- Building materials suppliers who provide the basic materials for construction such as lumber, cement and bricks;
- Machinery manufacturers who provide the heavy equipment used in construction such as cranes, graders and bulldozers;
- Building product component manufacturers who provide the subsystems (complex products) such as air quality systems, elevators, heating systems, windows and cladding;
- Sub-assemblers (trade speciality and installers) who bring together components and material to create such sub-systems;
- Developers and facility assemblers (or general contractors) who initiate new projects and co-ordinate the overall assembly;
- Facility/building operators and management who manage property services and maintenance;
- Facilitators and providers of knowledge/information such as scientists, architects, designers, engineers, evaluators, information services, professional associations, education and training providers;
- Providers of complementary goods and services such as transportation, distribution, cleaning, demolition and disposal;
- Institutional environment actors who provide the general framework conditions of the business environment such as the physical and

communication infrastructure, financial institutions and business/trade general labour regulations and standards.

The above list provides a basic typology of construction related activities; some of these actors may be suppliers or clients of others in the production process and specific firms can be involved simultaneously in several of above activities. Some larger firms offer a vertically integrated range of services from design, through manufacturing of some products or components, to building and operation of facilities.

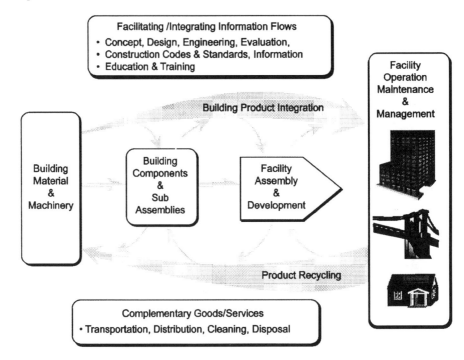

Figure 2.5 Key agents, major types of interactions and framework conditions in the construction sector

Source: Manseau (1998).

The ultimate client, the owner of the constructed facility, can influence the innovative behaviour of various actors by identifying specific novel requirements to be supplied by the developers, building product suppliers, contractors or operators or through the general expectation of high-quality and viability, over the life of the project.

2.3 Major Changes, Trends and Impact

Most studies of innovation have focused on rapidly changing sectors of the economy, such as microelectronics, biotechnology, information technology, software and business services, etc. Construction has been seen as a mature industry where changes are developing slowly. However, recent studies have shown that major and rapid changes are occurring in construction, but they are usually embodied in complex products or processes, and thus more difficult to observe.

Construction has a longer history than most industries and has been considered a backward industry, as it remains highly labour-intensive and productivity growth has been slower, as compared to many other sectors. However, despite the reputation of a stable industry, major forces are changing how the construction industry does business. Gann (1994) stressed five areas of major technological innovation that have had a considerable impact on construction products and processes:

- The use of information technology in the construction process;
- The use of information technology in buildings – 'intelligent buildings';
- Mechanization of construction activities;
- Prefabrication;
- New materials.

These changes are, however, difficult to observe and measure. They are embodied in larger systems or products, and often impact on the organizational arrangement between the different suppliers, trade specialties and general contractors. Organizational innovations are still not well captured by quantitative data.

Table 2.3 summarizes major trends affecting the construction industry and their impact on changing traditional roles and scope of the industry, as presented in an international study (Manseau & Seaden, 2001) and discussed in a Symposium in June 2001 (NRCC and CIB, 2001).

Integrated products and services

The roles of stakeholders have undertaken major changes in recent years. Manufacturers of building products and components are offering a variety of products and services that are affecting the role of specialty trades. Some products are so easy to assemble, such as door and window kits, lighting systems, kitchen cupboards, that no more speciality trade contractors are required. Some other products are so complex and customized, such as elevators, security systems, heating and air-conditioning systems, that these products are often installed and maintained by manufacturers themselves.

Engineering firms are becoming much more involved in the operation and maintenance of facilities and infrastructures. We have also seen a number of general contractors who now own and operate a large portfolio of built assets. Are they still contractors or in the real estate industry?

Major changes are also occurring from the demand side. As customers ask for complex and highly customized buildings, contractors and designers have to work closer together; the design-build contractual arrangement has become a preferred approach from the traditional division between these two steps. Again in this case, the role of general contractors has changed and is becoming integrated with consulting engineers and architects.

Information and communication technologies – Technology systems such as GPS and EDI, computer networks and design-monitoring systems (CAD and cost-schedule planning-monitoring systems) are already beginning to affect construction in a variety of ways, and their impact is expected to continue to grow. Design can be changed or completed while the building is under construction – providing significant flexibility with regards to client changing/evolving needs. More specialized facilities are needed as well as improved building controls and automation to reduce operating costs. ICT enables and enforces links between various players more than before, increased abilities in defining customer needs and in providing customized solutions; it has also facilitated development of internal standards (Uusikyla et al., 2003).

Attract and retain young professionals and skilled workers

These changes have occurred without a renewal of the workforce. An aging workforce, combined with the difficulty of recruiting youth was identified as a key issue in the International Symposium (NRCC and CIB, 2001). Construction is seen as non-attractive and too traditional for young professionals. Otherwise, knowledge intensive jobs are growing in some related-construction sectors that are off-site, such as in manufacturing of complex building components (heating, ventilation and air-conditioned systems, prefabrication, fire protection and security systems), complex project management and facility management of 'intelligent buildings'.

Table 2.3 Major changes occurring in the construction industry

Major changes	Impact
- Integrated products and services	- 'Pure' contractors are losing business as manufacturers are offering installation and after sale services for their building components, and as some integrated firms are offering complete services from design to build to operate buildings and infrastructure.
- Information and communication technologies (ICT)	- ICT facilitates monitoring of complex projects and adjustments during construction; ICT in buildings are facilitating operations and maintenance.
- Lack of skilled workers	- Knowledge intensive jobs are growing in computerized manufacturing of complex building components, in project management and in facility management.
- International agreements for codes and standards	- Increasing international trade and competition.
- New contracting arrangements	- Long-term public-private agreements are being developed for constructing and operating public facilities.
- Climate change and environmental issues	- Increasing pressure for reducing resource usage and waste, as well as for community consultations in major investment decisions.
- Life cycle costs	- Becoming an important criteria for customer – need to be better supported; repair and rehabilitation technologies are growing in profile.

International agreements for codes and standards

Construction has often been viewed as a local industry, but increasingly it is becoming national and international, sharing similar professional standards and building codes. Trade liberalisation under various international agreements (AIT, NAFTA, GATT, European Union and APEC) has increased the export potential of construction firms and materials suppliers, particularly those operating in shrinking domestic markets. Local regulation and product standards still impede market entry, but the trend is from national to internationally harmonized standards, codes and products, although it still remains so far incomplete. An important challenge is the development of objective-based building codes, which allows for more creative and integrative solutions to construction.

New contracting arrangements

New contracting practices are being adopted, including design-build, build-own-operate-transfer and public-private arrangements (PPP) to finance, build and operate public facilities/utilities. These new long-term public/private contracts are also called 'concessions', and they are becoming widely used in European countries, particularly for highways (Campagnac, 2001 CRIC, 2002: Molenaar, 2002).

From the demand side is the increasing importance of leasing spaces (office, industrial, commercial as well as residential). Constructing and operating facilities are being delegated to expert firms and professionals. Facilities management experts are less interested in developing tight prescriptive specifications and low-bid selection criteria for construction, but rather they seek 'best value' selection criteria that include flexibility of use, adaptability, and life cycle costs.

Climate change and environmental issues

Efforts to improve energy efficiency must be achieved without compromising the indoor environment. Pressure will increase to reduce land and primary resource usage and to recycle building waste. In addition, new regulatory measures and environmental assessment requirements are expected to increase the cost, complexity and time required for construction.

Life cycle costs and building longevity

The longevity and durability of buildings are becoming increasingly important, in part to meet environmental pressures for reducing waste and recycling, but also as a results of an increase awareness of life-long operating and maintenance costs. Consequently, repair and rehabilitation technologies will gain in profile.

Globalization of the construction industry

International construction activities are increasing, especially for very large construction projects. Although still limited to a small share of all construction activities as well as to a few international companies, these major projects appears to expand in high growth markets, such as the emerging or new industrializing countries. The top 20 international contractors accounted for a total of US$ 131 billion in revenue in 2001, with almost 60 per cent coming from international operations (Engineering News-Records, 2003).

Major mergers and acquisitions (M&A) have significantly increased over the last few years. De Valence (2002) noted that M&A averaged four per year from 1996 to 1998 then accelerated to about 12 per year during the 1999-2001 period. The largest international contractors are pursuing both operations growth and geographic diversification.

With the development of trade agreements (NAFTA, APEC, European Union, etc.), internationalisation is also growing in building products and financial services. Brochner (2002) stressed two major forces that have had a strong impact on integration and internationalisation: the development of information and communication technologies, and the widespread deregulation, in particular the liberalization of financial markets. These two forces imply an easier flow of information and global financial markets. The author also predicts an increased pressure for releasing legal restrictions on workforce flows.

2.4 Key Drivers of Change and Innovation

As nothing is simple in the construction industry, identifying key factors of change also proves challenging. Some authors have stressed that innovation and changes are often introduced by manufacturers of building products or equipment (Arditi and Kale, 1997). Some others suggested that merchants and building products wholesalers have a significant influence on adoption of new products by small contractors and specialty trades in construction (Agapiou and Flanagan, 1998). Bernstein and Lemer (1996) have stressed the importance of customers to drive innovation.

On the contrary, Nam and Tatum (1992) considered the concept that clients drive innovation a myth. Rather, they suggested (1997) that the key innovation champion is more likely to be the construction project manager, who could be designer, contractor or owner. Slaughter (1993) identified on-site builders as key people in introducing innovations in construction. Innovation process appears to be very behavioural in nature. Sexton and Barrett (2003) showed that owner's competencies and attitudes for developing innovations are key characteristics of innovative for small construction firms. Ling (2003) extended these individual behavioural factors to some inter-personal ones, such as project team interest, flexibility within the group, targeted task groups, and involvement.

Professional associations and regulators have also been identified as key players to facilitate innovation, acting as integrators and knowledge brokers (Winch, 1998).

Major clients have demonstrated a low interest in innovation, and most often seek to upgrade existing facilities, reduce operating costs or add capacity (Alistair et al., 2001). The Alistair et al study interviewed 59 senior personnel from major UK construction clients. Their main investment factor was to obtain the highest quality at the lowest initial cost. Although life-cycle cost was mentioned as important, respondents admitted that their organization did not use this measure.

However, recent studies have shown an emerging interest in major clients for life cycle costs and flexibility of use through years (Nkado, 2002; Statistics Canada and NRC, 2002). The need for flexible facilities is becoming important, as changes in business are more frequent. For example, office spaces are usually utilized by a number of different tenants through their life, as well as each tenant's requirements for facilities are changing through years (with new production processes and staffing). Clients are even more supportive for innovations when they can see actual favourable results (Dulaimi et al., 2003). From an analysis of barriers to innovation in the house-building sector, Ozani (2003) came to the conclusion that bringing customers into the entire process, from design and production to delivery and after-care can change the situation.

On the other hand, many studies have shown that innovation in construction appear to occur in reaction to external changes, instead of intrinsic drivers or stimulants (Pries and Janszen, 1995). Construction activities are booming when the overall economy is going well or when new kinds of buildings are required for production purposes or safety issues (Lansley, 1991). As noted in the previous Chapter, Bowley (1996) found that changes in products or processes have often been stimulated by the sudden difficulties of obtaining the traditional resources of steel, timber or labour. The author also stressed that builders are frequently unwilling to undertake work involving major innovations, because of the costs and risks that fall on the builder without capturing benefits. Bowley and later analysts repeatedly conclude that normal economic stimuli do not work for the construction sector; where there is a lack of Return on Investment (ROI) for innovators (Bowley, 1996). For example, Gann (2000) finds a changing construction industry responding to innovation rather than leading innovation development. In part this is due to structural barriers to innovation (see previous Chapter, Table 1.1).

The traditional 'design-bid-build' approach has left few possibilities for creativity and integration to the builders. They have rather focused on cost reduction and fragmentation of production (Dorée et al., 2003).

A key capability would be the ability of an industry actor to integrate the various aspects of providing a solution that not only respond to a problem but also elicit community-wide acceptance. Firms have to develop capabilities to deliver integrated systems solutions to clients and must manage learning processes from past experiences and from partnering firms. These are necessary to ensure the effective integration and use of technologies – both soft and hard competencies are thus required (Gann, 2000).

Finally, based on a series of international workshops undertaken in 2001 and 2002, Courtney and Winch (2003) concluded that technology was not a fundamental component of innovation or change in construction. The procurement systems and contractual arrangements would have had the largest influence. Design-build, public-private arrangements to finance, build and operate public facilities/utilities would have influenced the development of more integrated products and services, the life cycle cost approach and the use of more sophisticated information and communication technologies.

2.5 Toward a Service Industry

A number of changes occurring in construction have an impact on the end product of the construction process. The need for integrated products and services, the development of new contracting arrangements – particularly concessions, and the management of the entire life cycle of construction are all influencing the shift towards more services.

Managing life cycle costs is becoming a major economic activity. With the increasing importance of managing existing stock, repair and maintenance account for more than 45 per cent of all construction works in many industrialized countries (Carassus, 2003). This has a significant impact on a number of construction activities. Contractors are diversifying their activities, providing a series of services such as operating facilities, providing utilities (electricity, steam) or building maintenance services.

The importance of maintenance costs has also influenced new construction, with the need of assessing life cycle costs and services of new investment decision. Facilities managers are becoming more strategic in managing their portfolios, with an increased focus on services provided by these facilities.

A number of new services have emerged or have been significantly enhanced for addressing the need for managing the construction life cycle and more complex/sophisticated buildings. These new services include construction briefs, product designs, project management, facilities operations and maintenance, and project financing. Campagnac (2001) also observed a development of new professional specialties, as these services require new skills.

This trend towards a more service-oriented industry involves an increasing role for the client, flexible contractual arrangement and an emphasis on continuous communication, rather than on initial specifications. The change also has a significant impact on interactions between firms, increasing interdependencies and the need for a partnering approach.

2.6 Summary

The traditional method for measuring construction economic activities has been limited to production sites and has therefore blinded major industry changes that have occurred in the manufacturing side, such as the embodied technology in products and equipment, as well as major in the service side, with the increasing importance of operating and maintaining existing facilities.

The management of existing stock and of the entire life of constructions has significantly influenced a shift toward more services, as well as changes in the types of contractual arrangements in the industry towards longer term and public-private partnerships for major facilities.

References

Agapiou, A., Flanagan, R. and al. (1998), 'The Changing Role of Builders Merchants in the Construction Supply Chain', *Construction Management and Economics*, vol. 16, no. 2, pp. 351-361.

Alistair, G. F., Isack, G. and Isack, F. (2001), 'Client Drivers for Construction Projects: Implications for Standardization', *Engineering Construction and Architectural Management*, vol. 8, no. 1, pp. 45-58.

Arditi, D., Kale, S. and Tangar, M. (1997), 'Innovation in Construction Equipment and its Flow into the Construction industry', *Journal of Construction Engineering and Management*, vol. 123, no. 2 pp. 371-378.

Bernstein, H. M. and Lemer, A. C. (1996), 'Solving the Innovation Puzzle – Challenges Facing the US Design & Construction Industry', American Society of Civil Engineers, pp. 87-107.

Bowley, M. (1996), *The British Building Industry*, Cambridge University Press, London, particularly Chapter VI – 'The Development of Modern Building Structures and Industrial Change-Conclusions', pp. 169-179.

Brochner, J. (2002), 'Building Economics and Facilities Management: Knowledge and Incentives', Plenary Session, in Ben Obinero Uwakweh and Issam A. Minkarah (eds), *The Organization and Management of Construction – 10th International Symposium – Construction Innovation and Global Competitiveness*, University of Cincinnati, USA.

Campagnac, E. (2001), 'La « commande » comme nouveau marché de services : crise ou renouveau du professionnalisme ? Les leçons de l'expérience britannique', *Espaces et Sociétés : Projet urbain, maîtrise d'ouvrage, commande*, vol. 105-106, no. 2-3, Paris, pp. 17-55.

Carassus, Jean (1999), 'The Economic Analysis of the Construction Industry'. *Cahiers du CSTB*, vol. 405, Paris, France.

Carassus, J. (2002), 'Des flux au stock, de la production à la gestion, de l'ouvrage au service : mutation de la construction', in *Cahiers du CSTB*, vol. 3402, pp. 116-121.

Carassus, J. (ed.), (2003), *Construction Sector System: A New Framework for the 21st Century – A Comparative International Study*, Book submitted for publication.

CRIC (Centre for Research on Innovation & Competition – University of Manchester). (2002), 'Innovation in the Service Sector – Analysis of data collected under the Community Innovation Survey (CIS-2)', CRIC Working Paper No. 11.

De Valence, G. (2002), 'Globalisation and Changes in Ownership of the Building and Construction Industry', in Ben Obinero Uwakweh and Issam A. Minkarah (eds), *The Organization and Management of Construction – 10th International Symposium – Construction Innovation and Global Competitiveness*, CRC Press, pp. 556-571.

Dorée, A., Holmen, E. and Caerteling, J. (2003), 'Co-operation and Competition in the Construction Industry of the Netherlands', Working Paper, University of Twente. Netherlands.

Dulaimi, M. F., Ling, F.Y.Y. and Bajracharya, A. (2003), 'Organizational Motivation and Inter-organizational Interaction in Construction Innovation in Singapore', *Construction Management and Economics*, vol. 21, no.3, pp. 307-318.

Engineering New-Record (2003), *2002 ENR top 225 international contractors*, http://enr.construction.com.

Gann, David M. (1994), 'Innovation in the Construction Sector', Chap. 15 in Dodgson, Mark and Rothwell, Roy (eds), *The Handbook of Industrial Innovation*, Edward Elgar, London. pp. 202-212.

Gann, David M. (2000), *Building Innovation – Complex Constructs in a Changing World*, London: Thomas Telford.

Lansley, P. R. (1991), 'Organizational Innovation and Development', Chapter 4 in Stocks, R. (ed), *Competitive Advantage in Construction*, Reed International Books, Oxford, UK, pp. 128-138.

Ling, F. Y. Y. (2003), 'Managing the Implementation of Construction Innovations', *Construction Management and Economics*, vol. 21, no. 6, pp. 635-649.

Manseau, A. (1998), 'Who cares About the Overall Industry Innovativeness?', *Building Research & Information*, vol. 26, no. 4, pp. 241-245.

Manseau, A. and Seaden, G. (eds) (2001), *Innovation in Construction – An International Review of Public Policies*, Taylor & Francis – Spon Press, London.

Molenaar, K. R. (2002), 'Innovative Contract Administration: A Report of the European Contract Administration Scan Tour', in Ben Obinero Uwakweh and Issam A. Minkarah (eds), *The Organization and Management of Construction – 10th International Symposium – Construction Innovation and Global Competitiveness*, CRC Press, Boca Raton Florida. pp. 184-197.

Nam, C.H. and Tatum, C.B. (1992), 'Strategies for Technology Push: Lessons from Construction Innovations', *Journal of Construction Engineering and Management*, vol. 118, no. 3, pp. 507-524.

Nam, C.H. and Tatum, C.B. (1997), 'Leaders and Champions for Construction Innovation', *Construction Management and Economics*, vol. 15, no. 3, pp. 259-270.

National Research Council of Canada (NRCC) and the International Council for Research and Innovation in Building and Construction (CIB) (2001), *Construction Innovation: Opportunities for Better Value and Profitability*. Symposium, Ottawa (Canada), June 6-15, 2001.

Nkado, R. and Mbachu, J. (2002), 'Comparative Analysis of the Performance of Built Environment Professionals in Satisfying Clients' Needs and Requirements' in Ben Obinero Uwakweh and Issam A. Minkarah (eds), *The Organization and Management of Construction – 10th International Symposium – Construction Innovation and Global Competitiveness*, CRC Press, Boca Raton Florida. pp. 408-425.

OECD (2002), *Annual Accounts – Main Aggregates*. OECD, Paris.

OECD (2002), *STAN Industry Structural Analysis*. OECD, Paris.

Ozaki, R. (2003), 'Customer-focused Approaches to Innovation in House-building', *Construction Management and Economics*, vol. 21, no. 6, pp. 557-564.

Pries, F. and Janszen, F. (1995), 'Innovation in the Construction Industry: The Dominant Role of the Environment', *Construction Management and Economics*, vol. 13, no. 1, pp. 43-51.

Sexton, M. and Barrett, P. (2003), 'Appropriate Innovation in Small Construction Firms', *Construction Management and Economics*, vol. 21, no. 6, pp. 623-633.

Slaughter, S. E. (1993), 'Builders as Sources of Construction Innovation', *Journal of Construction Engineering and Management*, vol. 119, no. 3, pp. 532-549.

Statistics Canada and National Research Council Canada (2002), 'Improvement and Innovation in Construction Investments Survey 2000', Statistics Canada Questionnaire, Ottawa.

Uusikyla, P., Valovirta, V., Karinen, R. Abel, E. and Froese, T. (2003), 'Towards a Competitive Cluster – An Evaluation of Real Estate and Construction Technology Programmes', *Technology Programme Report* TEKES Helsinki, Finland June.

Winch, G. (1998), 'Zephyrs of Creative Destruction: Understanding the Management of Innovation in Construction', *Building Research & Information*, vol. 26, no. 4, pp. 268-279.

Chapter 3

Redefining Innovation

André Manseau

It is difficult to apply the current models of industry innovation and changes outlined in Chapter 2 to construction. Major changes have occurred in this industry, as stressed in the previous chapter, but traditional measures of innovation, such as R&D expenditures, number of patents, or workers' skill levels, are rather low and innovation seems to not be well captured by these indicators.

The difficulty of assessing innovation has also been stressed in the Service sector (CRIC, 2002). Almost all of our understanding of innovation derives from studies of manufacturing. The construction production process is quite different from manufacturing, as it involves intensive interactions among a number of different players. Organizational practices, and particularly co-operative arrangements between different firms and organizations, are critical to the production system. Despite recent important efforts to measure organizational innovations in many OECD countries, it is still difficult to assess. These innovative activities are generally pervasive throughout an entire firm and are also embedded in its interactions with key partners. Firms do not capture and maintain records of these kinds of innovations.

It is now generally accepted that innovative activity takes place within a 'system of innovation' (Amable et al, 1997; Lundvall, 1992; OECD, 1996 and 1997b). However, this concept can be approached from the perspective of technology clusters (Foray and Freeman, 1992; Van de Ven, 1993)), of a firm interacting with its business networks and within its competitive environment (Porter, 1998), of a region or geographic cluster (coming from a broad literature on urban development, see for instance Jacobs, 1969) or as a broader social process that we will further develop in section 3.6.

While full analysis of systems of innovation is clearly outside the scope of this volume, we will briefly review the different models that have been gradually introduced. First, the linear models, such as the Technology-Push or the Market-Pull, are still commonly used to analyze innovation. Then, three non-linear models of the innovation system approach are presented. Each model brings its specific approach for defining and measuring innovation. Comments as to the relevance of various models to the innovation in the construction industry are made. Construction, described as a complex industry in the previous chapters, has many facets and elements of each model could be applied for a comprehensive perspective.[1]

3.1 Technology Push or Science-based Model

The Second World War provided a unique implementation opportunity for science and technology. For the first time ever, with the total commitment to the victory by all warring parties and with the mobilisation of national resources towards the war effort, new technology-based weaponry was rapidly developed from scientific principles known technologies to operating products. The atomic bomb, the radar, the jet-propelled aircraft or mass production of Liberty ships were all developed through close collaboration of scientists, engineers and production personnel.

In the post-war years, this linear model of innovation became the reference standard that is now deeply entrenched in various policy instruments of governments as well as in the public perception of innovation. It assumed that basic (pure) research followed by applied research led from experimental development to new products or processes. Thus, the propensity of an industrial sector or a firm to innovate could be measured by its research intensity using several proxy indicators such as expenditure on R&D, citation analysis, or the level of education of research professionals. With this model, innovation is focused on formal R&D activities. Internationally (with the exception of Japan) the level of R&D efforts in the construction industry has been very low, which has led believers in the Technology Push model to conclude that it is a non-innovative industry. However, an R&D focus has proven to be a limited perspective for assessing innovation in construction, as well as in many other sectors. Many innovations in construction occur at boundaries between speciality trades and in organizational innovations, which are not captured by formal R&D measures (Gann, 1994). Recent findings and OECD surveys have shown that R&D constitutes only one of the sources of industrial innovation among many other alternatives (OECD, 1997a). Intensity of knowledge flows, levels of co-operation and effectiveness of technology diffusion may be more significant, particularly for industries such as construction (OECD, 1998).

3.2 Market Pull Model

Studies of highly successful and profitable firms (Cooper, 1998) indicate that they have very close ties to their customer bases. Their innovation activity linked to actual market opportunities. The challenge of arriving at innovative products/processes that are truly competitive (i.e. meet clients' needs, have superior quality, reduce costs and present visible benefits) is met by tapping into the vast pools of existing knowledge. Close contacts are established with various knowledge sources and feedback loops are extensively used at different stages of development.

This model suggests that clients and customers are the key drivers for innovation. The focus is on the demand side, and therefore innovation is assessed by measuring the number and scope of new products introduced.

As mentioned in section 2.4, clients are often indicated as an important driver of innovation by construction firms (Bernstein and Lemer, 1996; Manseau & Seaden, 2001). However, it appears that innovative clients are rather rare. Alistair et al (2001) showed that major clients were rather focused on the lowest initial cost and standardized products and processes. Although recent studies have shown an emerging interest for innovation by major clients (Nkado, 2002; Statistics Canada and NRCC, 2002), it is still rather limited to a few large clients.

Construction investments are usually of a high cost and carry significant risk. Risks include not only safety and the reliability of the construction itself, but also the difficulty of replacing or modifying a building structure once completed.. Furthermore, there is also the significant impact the building will exert on its surrounding community for many years.

Unsafe buildings or structures can have dramatic consequences on future tenants or neighbours. These effects are called 'externalities' in economics - as they affect people beyond those directly involved in a particular economic transaction. The construction industry is highly regulated through building codes and labour regulations. It is further restricted in cities or in areas of high population density.

The market pull approach is also less relevant or less directly effective in a regulated industry. Regulation constrains the freedom of clients to build regardless of risks or restrictions. Thus regulation is often identified as a barrier to innovation in this approach.

3.3 Firm-centred Knowledge Networks

The knowledge network approach is a macro/micro-economic model used as the theoretical basis of what is widely referred to as *The Oslo Manual* (OECD, 1997a) for the measurement of innovative activity. It casts the firm as the 'innovation dynamo' (where the economic benefits of innovation can be appropriated) at the centre of an enabling network of suppliers, competitors, clients, as well as educational, communication, financial and legislative resources. These actors and how they interact between each other in creating and exploiting innovation constitutes an innovation system.

This model has focused on the role of technology as a key source of innovation while recognizing the importance of organizational change as a potential source of innovation. However, in practice innovation is difficult to measure. This model highlights the significance of strategic intent in a firm and of its market performance due to technologically new or improved products or production process. Innovation is expected to be 'significant' and 'new to the firm' (but not necessarily to the particular industrial sector). It should bring enhanced performance benefits to the firm and/or the customer. The enabling network or system is taken into consideration by measuring technology dissemination, access to sources of information, internal/external barriers, and the potential impact of public policies (see also chapters 2 and 7).

Recent studies showed that distance was a principal factor in the ability of firms to interact in developing innovations. Local or regional 'Knowledge Clusters' are now recognized phenomena for facilitating innovation and regional economic development. In the Knowledge Economy – industry sectors strongly based on knowledge development – proximity is providing a competitive advantage, because innovation requires intensive interactions and relationships between partners (Porter, 1998).

Innovation is generally initiated for competitive reasons, to lower the unit cost of production and/or to obtain greater market share. However, the development of new products or services, that primarily appeal to customers' aesthetic or quality perception or personal taste and which can thus provide firms with a significant competitive advantage, are not considered innovative by the Oslo methodology.

The OECD model was devised following extensive studies of advanced manufacturing and high-technology sectors of the economy and it may not be fully applicable to other industrial groupings. So far, there has been very little in-depth analysis of various innovation framework factors related to the construction industry.

3.4 Production Systems

Implementation of new ideas happens through interaction between workers within organisational structures and constraints. Recent work (Amable et al, 1997) suggests that certain features of production systems may be particularly conducive to innovation while others tend to suppress it. The following factors and their consequences are considered as contributing to a positive climate for innovation:

- organizational flexibility (leading to) → rapid responses to changes and innovation;
- employee reward structure connected to corporate profitability → greater acceptance of technological changes;
- good and safe work environment → streamlined production systems;
- general policy of full-employment → enhanced investment in productivity;
- markets open to domestic/international competition → changes in sourcing of supplies, optimisation of work-processes, technological changes.

Redefining Innovation

On the other hand, the following factors are considered to have a detrimental effect on the innovation climate:

- work organised around strict functional definitions → slow and difficult response to technological changes;
- frequent lay-offs and technology-related unemployment → resistance to productivity enhancing initiatives;
- salaries based on market rates or collective agreements → little employee interest in quality or productivity;
- acceptance of unsafe work practices → obsolete equipment not replaced;
- relatively high level of general unemployment → investment in mass production;
- national/regional barriers to trade → reduced pressure to innovate.

This model, based to a great degree on the intensity of employees' involvement and participation suggests that the Taylor-type mass-production organisation is not conducive to innovation and that new, more flexible work structures need to evolve to encourage creation of new products or processes.

This methodological approach appears of particular relevance to the construction industry. Limited research has been done but it would appear that most of the detrimental factors listed are currently present in the industry. There are also a few of the positive elements, since a shortage of skilled labour has encouraged investment in new equipment and the open market has maintained a high level of competition. Some of these factors can be observed in the analysis of the Swedish construction industry McKinsey Global Institute (1995) found high costs and low productivity, mainly due to fragmented and inflexible work practices, low level of domestic competition and very strict, performance-driven building regulations.

This model focuses on process and measuring innovation with new manufacturing processes and, eventually new organizational processes. Organizational processes are considered very important in construction, as assembly methods and contracting arrangements are the core activities in this industry. However and despite efforts for measuring organizational innovation in many OECD countries, major challenges remain to be addressed. It is still difficult to measure activities such as organizational change, skill development, new management process and new marketing strategy. These activities are often pervasive through many activities, and are also embedded in interactions with partners.

Slaughter (2000) studied the production system in construction as a value-added chain with six implementation stages: Identification of alternatives; Evaluation of options; Commitment; Preparation; Use and modification and Post-use evaluation. Innovation occurs when members of all stages collaborate throughout the entire implementation process, learn from one step to another, and allow for returning to a previous stage (as design) when required.

3.5 Complex Product Systems (CoPS)

Another interesting model that has recently been applied to the construction sector is the Complex Product Systems (CoPS – see also Chapter 5). This model finds its origin in the project management literature and in the defense industry in the 1950s but has only been systematically applied to innovation processes in the early 1990s. The CoPS model involves close interactions and negotiations between many players and a system integrator who can rapidly change or modify major components of the project. The model is also characterized by its project orientation, involving a number of different firms that are required for producing a complex and usually unique or small batches product (Hobday, 1998). 'Systems integration' is arguably the core expertise of CoPS.

One of the first applications of this model to innovation processes has been in the flight simulation industry (Miller et al, 1995). Another early application, this one directly related to the construction sector, was with manufacturers and installers of home automation products (Tidd, 1995). Among key results, Tidd showed that the CoPS approach required managing across traditional product division boundaries and strong inter-firm linkages. For instance, firms in heating-ventilating-air conditioning systems have developed partnerships with firms in sectors where they had none or very low interactions before, such as micro-electronics and software, in order to manufacture as well as to install and provide services for the automated systems.

A series of case studies using this approach in construction have been published recently. Miller (2000) has applied the model to large engineering projects. Barlow (2000) applied CoPs to offshore oilfield construction projects conducted by BP and the Andrew Project. Partners created an alliance - working together with same objectives, gain-sharing mechanisms and an integrated management team. The study identified key success factors such as teamwork, trust and integrated communications. Successes and benefits appeared to be very significant for the partners; they shared 45 million Pounds Sterling in bonus, completed the project six months ahead of time and had no dispute.

Gann and Slater (2000) studied 30 organizations of design, engineering and construction firms involved in constructing complex products. The study stressed that a focus on project delivery is not enough for success. Firms need to learn from one project to another and develop their management processes. The authors concluded that many firms lacked central management capabilities. This model could become increasingly applicable to a large portion of the construction industry with the current trend towards integration and partnerships (design-build, Build-Own-Operate-Transfer (BOOT), manufacturers-installers-services, public-private partnerships; see previous chapters).

3.6 Social Process

Many studies (Bassala, 1998; Rosenberg, 1982; Niosi, 1995) have shown that the process of innovation is rarely an individualistic dynamic. The traditional myth is to consider the inventor or the entrepreneur as the driving force, creating technologies and developing the industry. These studies first focused on technical innovation, but quickly established intrinsic linkages between technical, organizational and social changes. Nelson and Winter (1982) described the innovation as a continuous and evolutionary process that occurs among many actors who interact collectively. Innovation is a social process.

With the globalization and the acceleration of technology changes, firms' strategies must deal with several partners in a mutual learning and complementary relationship. The individual, independent and conquering entrepreneur is changing toward a team-builder, creating consensus throughout an interactive process between multi-interdependent players. There are no isolated individuals, but rather social interactions, with emerging leaders, followers and some dominant actors.

Successful innovations are those that become a standard and reach wide diffusion and high utilization rates, building social networks around them (Foray and Freeman, 1992). Take for example a successful innovation such as the automobile, a number of social networks have been developed around this innovation, taking the form of social clubs, associations, journals, etc., and which have been instrumental in promoting this product.

The complementary aspect is also crucial and may be illustrated by a technology that fills a gap or a hole within an important or desired system. Hughes (1992) provided an application of this concept with the development of the electric system in the USA. The system has only had important growth after having resolved the problem of high voltage transmission, and transformer manufacturers become dominant in the industry, setting new standards. This means that in order to be successful an innovation must provide specific, complementary and desired contributions to a major market.

However, a complementary contribution or expertise should be continuously adapted because of partners or competition change. That involves a need for ongoing change and improvement in order to achieve a more efficient and adaptive way of working. Nelson and Winter (1982) described this dynamic as a successive adoption of better 'routines' (a routine can be a product, a process or an institution) that are evolutive or path dependent in relation to the learning capabilities of organizations. An organization is ever changing and searching for a better match (Teece, 1985) and building new 'consolidations' that are new production processes or new business arrangements with suppliers, partners or vendors (Rothwell and Gardiner, 1989). We may notice that the meaning of 'consolidations' used by Rothwell and Gardiner is close to the one of 'routines' developed by Nelson and Winter.

A central implication of the social approach in innovation is a deep-cutting rejection of 'maximization'. Human action and interaction needs to be understood as largely the result of shared habits and thought (Nelson & Nelson, 2002). Humans do not calculate everything (although they may behave as if they do). Any significant change is risky, and each economic actor (individual or firm) has a limited set of alternatives that they can access and which they can master in their particular context. This is why change or innovation is rather progressive and based on past experience and cumulative learning. Innovation tends to be the result of the cumulative contributions of many parties, linked to a number of other innovations that have been developed over many years.

Technology embodies assumptions and ideas about social arrangements. Innovation is closely related to a process of changing networks of social relations. Resistance is manifested in fears about changes in established networks of social relations (Woolgar, 1998). Competitive advantage depends upon technological as well as relational tools. Effective structure or network of relationships provides firms with lower transaction costs, tacit knowledge sharing and a high level of trust with partners and clients, which is instrumental in business development and for new market penetration.

3.7 Facets of Innovation in Construction

In reviewing various models presented in this section, we observed the difficulty of describing the complexity and the multi-dimensional aspects of innovation activities. Applying such models to the construction industry makes the task even more difficult since none of them appears to have the overall 'best-fit' to the specific characteristics of this economic sector.

The multi-dimensional aspects of innovation in construction can be illustrated by the development of improved energy efficiency housing. Energy efficiency has required the development of a number of new products, such like isolative material, thermo-glass, better heating, ventilating and air-conditioning systems (including control systems), reduction of material with toxic emission, etc. as well as a number of new construction methods and designs for reducing thermal transfer and vapour condensation, and for a better use of both passive (sunlight, water reservoir) and active sources of energy. The overall innovation process has been very complex, integrating innovations from manufacturers as well as being influenced by new energy efficiency regulations or codes, and by users' behaviours (life cycle and energy cost awareness, resistance to close environment, etc.).

However, we observed that these models are mutually complementary, representing various perspectives and different drivers of innovation. Some models focus on specific drivers, such as the science-technology capability of a firm, its organizational context or its clients' needs. Other models have a broader macro-view, where the interactions between a number of different players and the nature of connecting networks are taken into consideration.

It could be argued that during the post-war 'golden period' of construction, between the 1950s and 1970s, a synchronous interaction between different players and congruence of public and private objectives prevailed in many countries. This period is also of importance because of the intensity of construction innovation at the time. Since then, the economic environment, the characteristics of the industry and the public policy context have significantly evolved. Although there have been some recent efforts to develop national goals or strategies in construction, such as those in the UK, USA and Australia, their implementation and tangible results are still expected.[2]

As seen in Chapter 2, a number of changes and innovations have recently occurred in construction and no single player can be identified as the sole driver of change. Construction activities involve a complex set of interconnected players, therefore innovation occurs through complex and cumulative interactions amongst many players. The trend toward more of a service-type industry in construction has also a significant impact on innovation processes.

Because of its complexity, innovation processes are still poorly understood in construction (Seaden, 2002). In addition to component suppliers, architects, specialist consultants, trade contractors, principal contractor or engineering firm, client and regulators interacting all together throughout the construction process, we also have 'innovation brokers' that facilitate flows of information and knowledge between players (see Chapter 5), as well as a wider stakeholder community interested in harmonizing construction with the natural and social environment.

There are a number of reasons that could explain the need for intensive interactions between the different players. Firstly, any new construction, if it is not an identical copy of a previous project, requires a different set of components that have to fit perfectly in a complex arrangement. Each contributor or participant needs to coordinate with all others in order to ensure a successful assembly of this unique work. The building process is so complex that it usually costs as much, if not even more, that the total cost of components. Secondly, the final product cannot be pre-evaluated or tested, such as beta products in manufacturing. From a recent Canadian survey (Statistics Canada and NRCC, 2002), 85 per cent of major clients conduct prior assessments of their suppliers (principal contractors, architect or engineer), and 78 per cent contract with pre-qualified suppliers.

Developing effective interactions is important as it stimulates innovation more than the development of new products. Almost all innovative contractors (93 per cent) indicated that innovation in 'business-to-business arrangements' provided them with a significant competitive advantage followed by 'innovative business strategy' (see Table 1), then by new equipment and communication technology. Introducing new software or a new product impacted fewer firms, 46 per cent and 39 per cent respectively. According to these responses, software or new products give less than half as much advantage or benefit in comparison with changes in relations to other firms, such as the way in which subcontracting is managed. Moreover, client relationship is the most important business success factor for engineering firms, for building component manufacturers as well as for contractors (Statistics Canada, 2001).

Table 3.1   Sources of competitive advantage for contractors (Statistics Canada, 2001)

Innovation	**Percentage of Respondents**
Business-to-business arrangements	93
Innovative business strategy	84
New equipment	71
Communication technology	61
New software	46
A new product	39

Construction is clearly a multi-faceted industry where elements from each innovation model can be applied. Some innovations are introduced by new products that might follow a technology push model. Some others, particularly initiated from major clients who are increasingly aware of life cycle costs and flexible usability of their construction investment, can be of a market pull model. However, we have seen that the most important innovations are of the organizational type, and therefore elements of the production or complex product system models are certainly of a great help. Finally, a broader approach, like the social process model, would help in understanding paths of innovations through years and across countries (and cultures), the increasing role of various communities in major construction decision, as well as the evolution and development of regulations and standards.

3.8 Summary

Innovation in construction is a complex process involving a number of interrelated components and actors. Technology is usually embodied in a number of different materials, products or equipment. New methods for construction are progressively introduced, often influenced by a change in regulation or code and not seen as innovations. One of the most visible aspects of innovation is in business-to-business arrangements however, this is still poorly documented. Business-to-community interactions, including public sector organizations, appear to become increasingly important in developing innovations in construction as well as for driving changes.

Notes

1. These sections are based on previous work of the author (as can be found in Manseau and Seaden; 2001), which has been revisited and expanded.
2. In Australia, the Federal Government released an Action Plan to enhance the construction sector's performance in May 1999, based on a report completed by the AEGIS (Australian Expert Group in Industry Studies): 'Mapping the Building and Construction Product System in Australia'. In 1994, a committee of 14 federal agencies in the USA, part of the National Science and Technology Council, and in collaboration of the industry developed a plan for achieving seven national goals in the construction industry over the next decade. Finally, in the UK the agenda for change in construction set by the Lathan Report in 1994 was reinforced by the Egan Report – Construction Task Force in 1998.

References

Alistair, G. F., Gibb and Fran Isak (2001), 'Client Drivers for Construction Projects: Implications for Standardization', *Engineering Construction and Architectural Management*, vol. 8, no. 1, pp.45-58.

Amable, A., Barre, R. and Boyer, R. (1997), *Les systèmes d'innovation à l'ère de la Globalisation*, Ed. Economica, Paris.

Barlow, J. (2000), 'Innovation and Learning in Complex Offshore Construction Projects', *Research Policy*, vol. 29, no. 7/8 pp. 973-989.

Bassala, G. (1988), *The Evolution of Technology*, Cambridge University Press, Cambridge.

Bernstein, H. M. and Lemer, A. C. (1996), 'Solving the Innovation Puzzle', Chapter 6 in ASCE, *Challenges Facing the US Design & Construction Industry*, American Society of Civil Engineers, Washington. pp. 87-107.

Cooper, R.G. (1998), *Product Leadership: Creating and Launching Superior New Products*, Addison-Wesley, Reading, USA.

CRIC (Centre for Research on Innovation & Competition – University of Manchester). (2002), 'Innovation in the Service Sector – Analysis of data collected under the Community Innovation Survey (CIS-2)', CRIC Working Paper No. 11.

Foray, D., and Freeman, C. (1992), *Technologie et Richesse des Nations*, Ed. Economica Paris.

Gann, D. M. (1994), 'Innovation in the Construction Sector', Chap. 15 in Dodgson, Mark and Rothwell, Roy (eds), *The Handbook of Industrial Innovation*, Edward Elgar, London. pp. 202-212.

Gann, D. M. and Slater, A. J. (2000a), 'Innovation in Project-based, Service-Enhanced Firms: The Construction of Complex Products and Systems', *Research Policy*, vol. 29, no. 7/8, pp. 955-973.

Gann, D. M. (2000b), *Building Innovation – Complex Constructs in a Changing World*, Thomas Telford London.

Hobday, M. (1998), 'Product Complexity, Innovation and Industrial Organisation', *Research Policy*, vol. 26, pp 689-710.

Hughes, T. P. (1992), 'The Dynamics of Technology Change: Salliants, Critical Problem, and Revolutions', in Dosi, G. Giannetti, R., Toninelli, P. A., (eds), *Technology and Enterprise in Historical Perspective*, Clarendon, London. pp. 97-118.

Jacobs, J. (1969), *The Economy of Cities*, Random House, New York.

Lundvall, B-A. (1992), 'Introduction' in Lundvall, B. (ed), *National Systems of Innovation: Towards a Theory of Innovation and Interactive Learning*, Pinter, London. pp.2-19.

Manseau, A. and Seaden, G. (eds) (2001), *Innovation in Construction – An International Review of Public Policies*, London: Taylor & Francis – Spon.

McKinsey Global Institute (1995), Sweden's Economic Performance. Private Communication.

Miller, R., Hobday, M., Lerous-Demers, T. and Olleros, X. (1995), 'Innovation in Complex Systems Industries: the Case of Flight Simulation', *Industrial and Corporate Change*, vol. 4, no.2, pp. 363-401.

Miller, R., Lessard, D.R., Michaud, P., and Floricel, S. (2000), *The Strategic Management of Large Engineering Projects: Shaping Institutions, Risks and Governance*, Cambridge: MIT Press, Cambridge, USA.

Nelson, R. R. and Winter, S. (1982), *An Evolutionary Theory of Economic Change*, Belknap Press of Havard University.

Nelson, R. R., and Nelson, K. (2002), 'Technology, Institutions, and Innovation systems', *Research Policy*, vol. 31, no. 2, pp. 265-272.

Niosi, J. (1995), *Vers L'innovation Flexible – Les Alliances Technologiques dans L'industrie Canadienne*, Presses de l'Université de Montréal, Montreal.

Nkado, R. and Mbachu, J. (2002), 'Comparative Analysis of the Performance of Built Environment Professionals in Satisfying Clients' Needs and Requirements' in Ben Obinero Uwakweh and Issam A. Minkarah (eds), *The Organization and Management of Construction – 10th International Symposium – Construction Innovation and Global Competitiveness*, CRC Press, Boca Raton Florida. pp. 408-425.

OECD (1996), *The Knowledge –based Economy*. OECD, Paris.

OECD/Eurostat (1997a), *Proposed Guidelines for Collecting and Interpreting Technological Innovation Data - Oslo Manual*, OECD, Paris.

OECD (1997b), *National Innovation Systems*, OECD, Paris

OECD (1998), *Science, Technology and Industry Outlook-Construction*, OECD, Paris

Porter, M. E. (1990), *The Competitive Advantage of Nations*, The Free Press, New York, USA.

Porter, M. E. (1998), 'Clusters and the New Economics of Competition', *Harvard Business Review*, vol. 76, no. 6, pp. 77-90.

Rosenberg, N. (1982), *Inside the Black Box – Technology and Economics*, Cambridge University Press.

Rothwell, R., and Gardiner, P. (1989), 'The Strategic Management of Re-innovation', *R&D Management*, vol. 19, no. 2, pp. 147-160.

Seaden, G. (2002), 'Changing more than R&D: Responding to the Fairclough Review', *Building Research and Information*, vol. 30, no. 5, pp.312-316.

Slaughter, E. S. (2000), 'Implementation of Construction Innovations', *Building Research & Information*, vol. 28, no. 1, pp. 2-17.

Statistics Canada and National Research Council Canada (2002), Improvement and Innovation in Construction Investments Survey 2000, Data, Statistics Canada, Ottawa.

Statistics Canada (2001), Innovation, advanced technologies and practices in the construction and related industries: national estimates, Science, Innovation and electronic Information Division, Working paper prepared by Anderson, Frances and Schaan, Susan, Catalogue No. 88F0006XIB2001004. Statistics Canada, Ottawa.

Teece, D. J. (1985), 'Applying Concepts of Economic Analysis to Strategic Management', in Penning and Associates, *Organizational Strategy and Change*, Josey Bass, New York.

Tidd, J. (1995), 'Development of Novel Products Through Intra-Organizational and Inter-Organizational Networks – The Case of Home Automation', *Journal of Product Innovation Management*, vol. 12, no. 4, pp. 307-322.

Van de Ven, A. H. (1993), 'Innovation and Industry Development: The case of Cochlear Implants', *Research on Technological Innovation, Management and Policy*, vol. 5, pp. 1-46.

Woolgar, S. (1998), 'A New Theory of Innovation?', *Prometheus*, vol. 16, no. 4, pp. 441-452.

Chapter 4

Measuring Innovation in Construction

Frances Anderson

4.1 Introduction

This chapter presents different approaches to the measurement of innovation in construction and examples of findings that use these different approaches. Statistical information collected on firms and other organizations usually relies on surveys that are sent to firms posing questions on their activities. As some national statistical agencies have listings of all firms in a particular country, a representative sample of firms can provide statistical information on the entire population of firms. The first section of the chapter will discuss the issue of how to define construction activities, and will be followed by sections describing two different measurement approaches to innovation: the 'firm as actor approach'; and the systems approach. A final section will present econometric analyses which that deal with the determinants of innovation.

4.2 What are Construction Activities?

What are construction activities? This is a critical first question in any attempt to measure construction activities. Traditionally, the construction industry has been defined by national statistical agencies as firms engaged in a set of activities that are directly related to the construction site. This approach can be called the Construction Industry Approach.

Thus, for example, construction industries included in the construction sector as defined by the North American Industry Classification System (NAICS) (1998b) are divided into two general categories: Prime Contracting and Trade Contracting (see Table 4.1). Prime Contracting includes those firms engaged in constructing complete works, whether buildings or engineering products, such as residential and non-residential building construction and engineering construction which involves construction projects such as highways, bridges, sewers, power and communications transmission lines and similar structures and works. These firms are normally called general contractors. Trade contracting includes firms engaged in one aspect of construction work that is common to different structures, but which generally require specialized skills or equipment. This set of industries is often referred to as the specialty trades and include firms involved in site preparation work, building structure work, building exterior and interior work and equipment installation. In specific terms, this includes activities such as excavating, carpentry, roofing, drywall, plaster work, plumbing, electrical work and air-conditioning to name only a few.[1]

Table 4.1 Industries classified in the construction sector by the North American Industrial Classification System 1997

.1 Prime Contracting	.2 Trade Contracting
• Land Subdivision and Land Development • Building Construction • Engineering Construction • Construction Management	• Site Preparation Work • Building Structure Work • Building Exterior Finishing Work • Building Interior Finishing Work • Building Equipment Installation • Other Special Trade Contracting

Source: Statistics Canada (1998b).

Most statistics that are currently produced by national statistics agencies on the construction industry are based on this type of definition of construction. A number of recent studies have challenged this Construction Industry Approach. They favour adopting a Construction Cluster Approach to defining construction activities. (See Den Hertog and Bourver, 2001; Vock, 2001, Dahl and Dalum, 2001; and Anderson and Manseau, 1999) These studies argue that an analysis based only on those industries that are found on the construction site presents a misleading view of the complexity of the construction production system. Critical to this approach is the inclusion of the suppliers to the construction site. Dahl and Dalum (2001), using the industry classification system based on the European Statistical Classification of Economic Activities in the European Community (NACE) system, provide a definition of the industry components of what they call the 'construction mega-cluster'. Their mega-cluster includes industries from the primary, the manufacturing and the service sectors. Den Hertog and Brouver (2001) include industries from the wholesale sector as well.

The Canadian case will be used as an example of the Construction Cluster Approach. This grouping or cluster is constructed on the basis of an existing classification system, the NAICS industry codes. Figure 4.1 is a graphic representation of the set of industries and organizations that comprise the construction cluster. The ones found below are similar but not identical to the construction cluster as defined in other studies for other countries.

At the heart of the construction cluster are firms involved in the value-added chain that leads to the final built structure. The value-added chain goes from the suppliers of raw material to the building components suppliers to the sub-system assemblers, and finally to the facilities assemblers. There is a set of additional firms which provide expertise, process equipment and financial support to the production system actors (professional services, suppliers of machinery, equipment, building owners and suppliers) and a set of public sector actors that provide the manpower, regulatory context and infrastructure support of the actors in the construction production system (educational/research institutions, government regulatory agencies, professional associations/unions, and infrastructure providers).

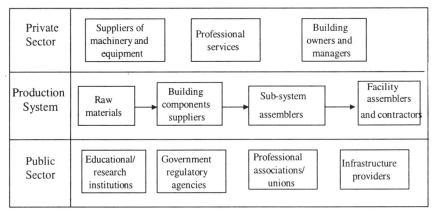

Figure 4.1   Actors in the construction cluster

Source: Anderson 2002.

The set of actors in Figure 4.1 represent one possible approach to defining the construction cluster. This set of actors could be expanded by taking into consideration the life-cycle of built structures. The three phases of the life cycle of a built structure are as follows:

- Phase 1: Constructing new built structures. The construction of new built structures involves the flow of products and services to the construction site where the built structure is assembled. These are described in Figure 4.1 as the production system;
- Phase 2: Maintaining, operating and managing built structures. Maintaining, operating and managing built structures is an important part of construction which also involves repair and renovation. Facilities owners and managers, both public and private, are important actors in this phase;
- Phase 3: Demolishing/recycling built structures. Finally, Phase 3 involves actors involve in the various aspects demolition, disposal, reuse of materials and site remediation.

Whereas Figure 4.1 represents the actors who are involved in Phase 1 of the life cycle, a different set of actors is involved in Phase 2 (Maintaining, operating and managing built structures) and Phase 3 (Demolishing/recycling built structures). The construction cluster could add these actors as well.

Returning to the question, 'What are construction activities?' posed at the beginning of this section, it can be seen that there are number of different ways to define construction activities. Whether or not one adopts the Construction Industry Approach or the Construction Cluster Approach, and whether one takes into consideration the life-cycle of the built structure, will have a great effect on the aggregate statistics that will be produced, both for making estimates of traditional indicators such as contribution to GDP, numbers of employees, or exports, as well as for measures of innovation.

4.3 The 'Firm as Actor' Approach

Research and Development expenditures are one component of innovation and are usually considered as 'input' indicators. The measurement of R&D has a long history dating back to the late 1960s. Prior to the 1990s, it was the principal indicator of a firm's activities in science and technology. The *Frascati Manual* (OECD 1994) sets guidelines for the collection of this type of statistic and is used by OECD countries to collect internationally comparable R&D statistics.

Using business R&D as the only measure of innovation, it has been noted by many observers that the construction industry has very low R&D expenditures relative to other industries. Statistics on R&D in construction generally use the limited set of industries as outlined in the Construction Industry Approach and, for the most part, when reported do not include the expenditures on construction R&D carried out by public organizations, be they universities or government research laboratories, nor of suppliers and other related firms.

By the end of the 1980s, the idea of a system of innovation arose and within this context, measures of innovation other than R&D arose. *The Oslo Manual* guidelines are currently used by a number of countries, for the collection of innovation statistics and for interpreting technological innovation data. Firms are identified as being innovators or non-innovators based on their responses to the following two questions posed by the Third Community Innovation Survey (2000):

- 'During the period 1998-2000, did your enterprise introduce any new or significantly improved products (goods or services) to the market?'
- 'During the period 1998-2000, has your enterprise introduced any new or significantly improved production process, including methods of supplying services and ways of delivering products?'

If the firm answers positively to either of these questions, it is considered to be an innovator and if it responds negatively to both questions it is considered to be a non-innovator. Innovation surveys typically ask a series of more detailed questions of the innovators. They include qualitative questions related to their innovative activities such as obstacles they confronted in carrying out innovative activities, the sources of information for innovation, collaborators on innovation projects, the impacts of innovation on the firm, public funding of innovation, R&D activities, and intellectual property practices.

Attempts have also been made to collect quantitative statistics as well. In the third CIS survey, for example, firms were asked to estimate expenditures for innovative activities in the categories listed in Table 4.2 (Eurostat 2000). There have been considerable difficulties in collecting this type of quantitative information because the standard accounting procedure of firms does not identify information using these categories.

What have been the results of the surveying construction activities using these innovation surveys? For the most part, there has been very little surveying of construction activities with surveys of innovation. Work has centered for the most part on the manufacturing and the service sector, with a couple of notable exceptions.

Cleff and Rudolph-Cleff (2001) found that 33 per cent of German construction firms undertook innovative activities in the years 1994-1996.[2] Relative to the manufacturing and service sector, construction is an industry with a below-average percentage of innovative firms. The share of innovators, however, was found to increase with company size. The share of innovators for companies with more than 200 employees is 59 per cent.

Figure 4.2 compares innovation expenditures by construction and manufacturing innovators. Whereas as for both sets of innovators a high percentage of innovators have expenditures on machinery and equipment, for the most part a significantly smaller percentage of construction have expenditures on the other types of innovation. This data confirms the results that have generally been obtained from R&D surveys (discussed above): construction firms spend less on both external R&D and intramural R&D than manufacturing firms. There is also a very significant difference between design of products with a few innovators in construction undertaking this type of expenditures. The one exception to the lower percentage found in construction is for expenditures on external know-how (expertise). Both industries have the same percentage of firms that have expenditures on the purchase of machinery and equipment.

The Statistics Canada Survey of Innovation 1999, consistent with the Construction Cluster Approach, surveyed the suppliers to construction.[3] All manufacturing firms were asked to respond to the following two questions:

- 'During the last three years, 1997-1999, did your firm offer products that *were incorporated into* buildings or other engineering works such as roads, dams, sewers, transmission lines and pipelines? Some examples of building products are windows, plaster board, bricks, concrete, heating and plumbing systems, roofing security systems, electrical systems and others;'
- 'During the last three years, 1997-1999, did your firm offer machinery, equipment or tools which were *used during the process* of constructing buildings and other engineering works such as roads, dams, bridges, sewers,

transmission line, and pipelines? Some examples of products used during construction are bulldozers, cranes, scaffolding, survey equipment and other.'(Statistics Canada, 1999b).

If the response was 'yes', a further question asked the respondent to indicate the percentage of total revenue from the sale of these products. Based on these questions, two types of suppliers to the construction were identified: 'Building product suppliers' and 'Equipment and machinery suppliers'.

A significant proportion of the manufacturing firms indicated that they were supplying building products to construction firms. Almost one quarter (23 per cent) of all manufacturing firms indicated that they supplied construction firms with building products. The highest percentage of building product supplier firms was found in Non-metallic Mineral Products (73 per cent of firms), Veneer, Plywood and Engineered Wood Products (58 per cent) and Petroleum and Coal Products (51 per cent). Six per cent of all manufacturing firms responded that they offered products used in the construction process (machinery, equipment and tools). The industries with the highest percentage of firms were found in the Agriculture, Construction, Mining and Industrial Machinery Industries (17 per cent), Electrical Equipment, Appliance and Component Manufacturing (16 per cent) and Non-Metallic Mineral Products (16 per cent). (Anderson, 2002) These industries, as is shown in Figure 4.3, have a relatively high percentage of firms that are innovative.

Sixty-one per cent of suppliers to construction were found to be product innovators compared to 68 per cent of all manufacturing firms. (Seaden, 2001) However, as can be seen in Figure 4.3, there is a significant difference between the different supplier industries, those that produce complex products such as machinery and equipment having a higher percentage of innovative products than those that produce standard products, such as wood or petroleum products. Seaden (2001) noted that construction suppliers ranked the client's ability to substitute as a competitive factor significantly higher than the manufacturing firms as a whole. He explains that this occurs as a result of the standardized/homogenized type of products that are being produced for the construction industry and the inability of construction suppliers to differentiate their products by added value. The study also found that construction product suppliers were less concerned with undertaking innovation to replace products being phased out than found in the manufacturing sector as a whole, which points to a more stable product life-cycle than is generally found in manufacturing.

Table 4.2 List of innovative activities in the 3rd CIS Innovation Survey

Intramural research and experimental development (R&D)	All creative work undertaken within your enterprise on a systematic basis in order to increase the stock of knowledge, and the use of this stock of knowledge to devise new applications, such as new and improved products (goods/services) and processes (including software research)
Acquisition of R&D (extramural R&D)	Same activities as above, but performed by other companies (including other enterprises within the group)
Acquisition of machinery and equipment	Advanced machinery, computer hardware specifically purchased to implement new or significantly improved products
Acquisition of other external knowledge	Purchase of rights to use patents and non-patented inventions, licenses, know-how, trademarks, software and other types of knowledge from others to use in your enterprise's innovations
Design, other preparations for production/deliveries	Procedures and technical preparations to realize the actual implementation of products (goods/services) and process innovations not covered elsewhere
Training	Internal or external training for your personnel directly aimed at the development and/or introduction of innovations
Market introduction of innovations	Internal or external marketing activities directly aimed at the market introduction of your enterprise's new or significantly improved products (goods/services)

Source: Eurostat (2000).

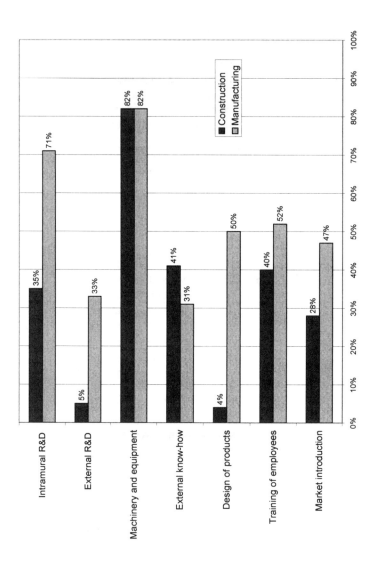

Figure 4.2 Percentage of firms with innovation expenditures, by type of expenditure

Source: Cleff and Rudolf-Cleff (2002).

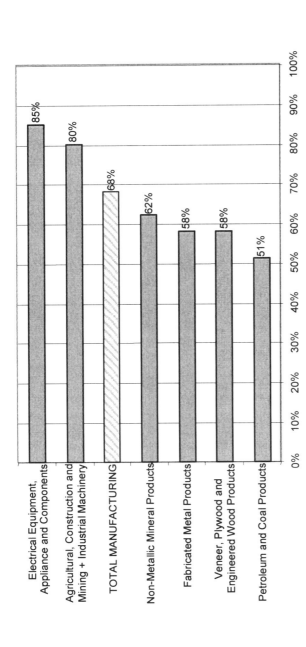

Figure 4.3 Percentage of product innovators in building products and construction equipment, machinery and tool industries, by selected manufacturing industries 1997-1999

Source: Statistics Canada Survey of Innovation (1999).

4.4 The Systems Approach

Traditionally, measures of science and technology, such as R&D and innovation, have produced indicators that measure the amount spent on particular organizations and the activities they were carrying out. This approach is what has been characterized above as the 'firm as the actor' approach. Such an approach provides the basis for an understanding of the characteristics of organizations; it does not, however, provide for an understanding of the interactions between scientific innovation on the one hand and the social, economic and political consequences on the other. Equally important is the measurement of the interactions between a growing number of actors in the innovation which are facilitated by increasing means of communications and transportation and by the increasing mobility of skilled workers.

The Science, Innovation and Electronic Information Division of Statistics Canada has developed a framework for the development of a statistical information system for science and technology (Statistics Canada, 1998a). This approach can be characterized as a Systems Approach. The framework is a basic operational instrument for the development of statistical information respecting the evolution of science and technology and its interaction with society, the economy and the political system of which it is a part. The framework sets out three types of activities in the science and technology (S&T) system that need to be considered when measuring science and technology related activities:

- The generation of knowledge;
- The transmission of S&T knowledge;
- The use of S&T knowledge.

It provides a framework for the firm innovation as well as for the other types of science and technology issues such as, for example, the impact of e-mail on social structures or the transmission of medical knowledge to medical practitioners. The application of the systems approach framework to the measurement of innovation in private sector firms is shown in Table 4.3.

Table 4.3 The application of the systemic approach framework to the measurement of innovation in private sector firms

Sets of Activities to be Measured	Quantitative Indicators of the Activity
Generation of S&T Knowledge	• Development of new or significantly improved products and processes
Transmission of S&T Knowledge	• Flow of innovative products and processes from generator to user
Use of S&T Knowledge	• Use and planned use of advanced technologies • Use and planned use of advanced practices

The European Community innovation surveys, which follow the Oslo manual, have tended to focus on issues on the generation of S&T knowledge, i.e. the development of new and significantly improved products and processes. They also focus on questions that provide details on the characteristics of the innovating or non-innovating firms. However, some of the questions have addressed system issues such as the relation between the responding firm, other firms and other types of organizations, for example, sources of information for innovation, collaborators for innovation projects, and public funders of innovation in the firms. The Canadian 1999 Survey of Innovation (Statistics Canada 1999b) has similar questions as well as additional questions that provide more details about the system in which the firm operates: the competitive environment of the firms, firm success factors, and the use of innovative products by other industries.

4.4.1. Advanced Technologies and Advanced Practices

Statistics Canada has developed another type of survey that addresses the question of the diffusion of advanced technologies and practices within an industry or a sector. It has carried out surveys on the use and planned use of advanced manufacturing technologies, and has recently carried out surveys of the use and planned use of biotechnologies; information and communication technologies; and knowledge management practices (McNiven 2002, Peterson 2001 and Earl 2002). At the heart of these surveys is a listing of specific sets of technologies that are developed by subject-matter experts. These technologies are judged by experts to be the most advanced technologies and practices that are being adopted by firms in a given set of industries. Advanced technology and practices surveys measure the extent that these technologies and practices are adopted by firms. Indicators of the diffusion of technologies and practices can be developed from these surveys.

In 1999, Statistics Canada carried out a survey of *Innovation, Advanced Technologies and Practices in the Construction and Related Industries*.[4] Listings of advanced technologies and advanced practices in the construction industry were developed. in collaboration with construction experts from the Institute for Research in Construction of the National Research Council of Canada and with industry representatives. The advanced technologies were grouped in the following categories: communication, on-site plant and equipment, materials, systems and design. Advanced practices were regrouped in the following categories: computerization, quality, organization and business. (Statistics Canada, 1999a).

The findings on the use and planned use of advanced technologies are presented in Figure 4.4 and the use and planned use of advanced practices is found in Figure 4.5. Significant variation was found in the use and planned used of both advanced technologies and advanced practices. Larger firms had significantly higher rates of adoption than did smaller firms. (Anderson and Schaan, 2001) An Australian study adopted a similar approach and had similar findings on the importance of size as a factor explaining adoption rates. (Manley, 2002) The Australian study surveyed not only contractors, but also surveyed clients (government agencies), consultants, and product suppliers, as well as extractive industries, equipment distributors and hiring firms.

The Oslo Manual makes the distinction between technological product and process (TPP) innovation and other types of innovation. It is recognized that TPP innovations involve a series of 'scientific, technological, organizational, financial and commercial activities' (Oslo Manual: 47). Thus it is clear, in the Oslo Manual, that organizational changes can be an important component of TPP innovation. Three types of purely organizational innovation are identified in the Oslo Manual: 1) the introduction of significantly changed organizational structures; 2) the implementation of advanced management techniques; and 3) the implementation of new or substantially changed corporate strategic orientations (Oslo Manual: 54).

Recognizing the difficulty of measuring non-technological innovation and, in particular, in measuring its impact on firm performance, the 1997 version of the Oslo Manual recommends that surveying agencies develop non-technological measures to be included in their surveys. In recent years, two different approaches have been developed. On the one hand, very general questions on non-technological innovation have been posed, as for example the CIS III which lists the three types of organizational innovation listed above and asks the respondent to indicate whether these activities have taken place in the previous three years.

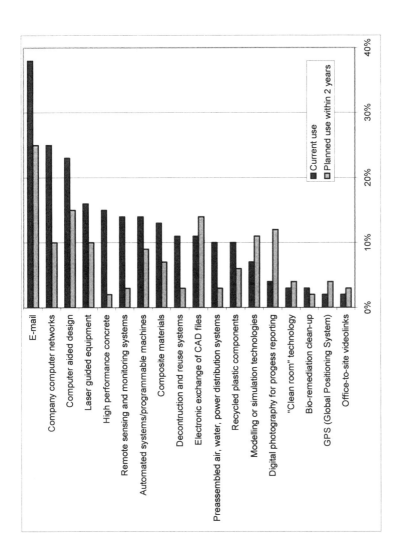

Figure 4.4 Percentage of businesses indicating use or planned use of advanced technologies within two years, all construction industries

Source: Anderson and Schaan (2001).

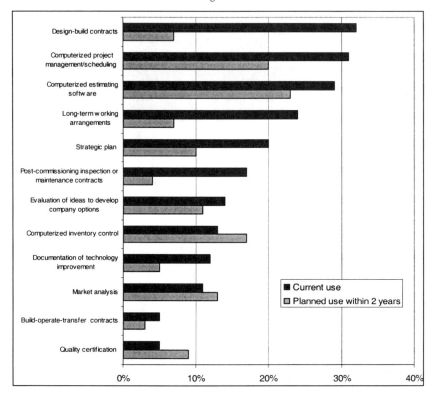

Figure 4.5 Percentage of businesses indicating current use or planned use of practices within two years, all construction industries

Source: Anderson and Schaan (2001).

The alternative approaches have been to ask the respondent if they are using or plan to use a listing of specific organizational practices. A good example of this approach is the knowledge practice surveys that have been piloted by the OECD. The Canadian and Australian approaches to organizational practices in the construction industry, which are discussed in this chapter, have adopted this approach.

4.4.2 Knowledge Flows

In the Systems Approach to innovation, it is important to measure not only the activities of individual firms, but also the transactions and interactions between firms and other organizations. Transactions and flows condition innovation activities and thus the measurement of such flows is critical to understanding innovative activities in construction. Different types of transactions and flows are possible and need to be explicitly examined. On the one hand, firms can simply buy products in the marketplace that incorporate innovative features (sometime characterized as 'embodied knowledge'). At the other extreme, user firms can be extensively involved with innovative producer firms, being involved in all phases of the innovative process, from design to prototype testing to final use. The receptor capability and knowledge capacity of firm is critical to its ability to adopt an innovative product, and the lack of receptor capability or knowledge capacity can result in the inability of a firm to adopt a given innovative product. Therefore, the following questions need to be addressed:

- What kinds of transactions and flows occur in construction?
- What kinds of transactions and flows are of relevance to innovative activities?
- What are the conditions that facilitate or prevent transactions and flows of relevance to innovative activities?

Construction activities can be characterized by the following types of transactions and interactions between channels through which knowledge information and products flow.

- *Processes related to product integration*: We have described above production processes that characterize construction. Along the chain of production, important interactions occur between the manufacturers of building and construction products and the assemblers and installers of these products, as components become part of sub-systems and sub-systems become part of the final built structure. The design of products and the system is an important area where the issues of product integration are worked out between actors.
- *Processes related to project organization and co-ordination*: Construction activities are organized and carried out within the organizational structure of projects, as opposed to the factory production system which characterizes the manufacturing system. Thus the intense interaction between actors on the

construction site or those involved in construction and repair is a precondition for carrying out work. Thus a very extensive network of partnering between actors characterizes construction activities.
- *The diffusion of technologies and practices*: Innovative products, materials, equipment and machinery flow into the construction site where the activities are characterized by assembly operations. The adoption of new products can give rise to a secondary innovative effect on sub-systems and systems or on the modification of assembly practices. Furthermore, much of the innovative business practices in construction as well flow onto the construction site from the service sector in the form of software programs or codified business practices. Examples of this are cost estimating software, project management software or ISO9000 quality control systems.
- *The flow of human resources*: The flow of trained people into construction is a critical component of innovation, especially the new graduates from technical schools, colleges and universities who bring with them the latest knowledge being generated.
- *Information flow*: The flow of information from various sources is a final important linkage.

The dynamics within the S&T system involve the flow of S&T knowledge from the site of generation to the site of use. This requires the means of transmission to get from one site to the other, and the capacity to transmit on the part of the knowledge generators and the capacity to search and absorb on the part of the knowledge users. Transmission should not been seen as a simple process of transfer or a handing-off of a new product from the supplier to the user. As was pointed out above, the user could be extensively involved in the innovation process itself and in the development of the product. Furthermore, a new product can require important changes to the user firm; for example, a firm that is faced with the adoption of a new software program to improve the efficiency of its operations might not have the skilled personnel to operate the software. Training might be required for all staff that would be using this new software. What is more, important changes in management practices might be required. This illustrates the importance of carefully considering the various obstacles to the transmission of the product. Equally important for the transmission of new products are issues of liability and reliability. There might be resistance to a new product because old products are 'tried and tested' and new product might appear too risky for the user. Facilitating actors can have a role to play in demonstrations and testing that assure product quality and reliability. Finally, there are facilitating actors who can be involved in the diffusion of information about new products. Trade fairs and exhibitions have been found to be very important vehicles to diffuse information about new products. Figure 4.6 describes the actors that can be involved in the transmission of new products.

Generation	Transmission	Use
MANUFACTURING PERFORMERS	──────────▶	CONSTRUCTION PERFORMERS
Suppliers of machinery and equipment Building products suppliers Suppliers of services	Direct supplier/user relations Facilitating actors	Sub-system assemblers Facility assemblers

Figure 4.6 Actors involved in the transmission of new products

In some cases, the products from other firms can result in new process innovation for construction performers. This is the case, for example, when machinery and equipment are bought by construction firms.

The German, Canadian and Australian surveys of construction have all contained questions on knowledge flows. As was mentioned above, the Oslo-type survey does measure knowledge flows with questions on sources of information and innovation collaboration. Concerning at the flow of information into the construction industry, Cleff and Rudolph-Cleff (2001) of construction innovators found that competitors have the highest significance followed by suppliers and clients. The study also found that fairs and conferences played a very important role for sources of information on innovation. With regard to research-based information from universities, larger firms were found to use this type of information more than the small or medium firms (SMEs). The authors stressed the importance of in-house innovative activities:

> It is obvious that SMEs co-operate on a lower level with institutions in the science sector such as universities and research institutions. The ability to adopt external technical knowledge from research institutions is conditional on in-house corporate potential. In-house innovative activities, expressed in R&D or innovation intensity, increases the likelihood of companies co-operating with universities (Cleff and Rudolf Cleff 2001, p. 207).

Collaboration is a more intensive form of transaction between organizations than is the flow of information. It involves joint activities and an active partnership carried between organizations. Cleff and Rudolph-Cleff (2001) found that the most important collaborators for innovative construction firms were clients followed by competitors and suppliers. Larger firms were found to collaborate extensively with partners in their own company group as well as with customers and consultancy firms, whereas small and medium firms collaborated mostly with suppliers or competitors. It was also found that if a small or medium-sized firm had its own in-house innovative activities, cooperation with universities was more likely and that SME subsidiaries of large companies were more likely to cooperate with universities than were independent medium-sized companies of equal size.

Figure 4.7 shows that there are important differences in the co-operation partnership between manufacturing and construction firms. Whereas both have similar percentages of collaboration with suppliers, clients and other firms in the same enterprise, construction firms collaborate significantly less with university/research institutes and consultancy enterprises and significantly more with their competitors. (Cleff and Rudolph-Cleff, 2001).

4.4.3 Business Strategies and Competitive Environment

The Australian and the Canadian surveys posed questions concerning the business strategies and the competitive environment of construction firms. The findings from the Australian study on the Queensland Road and Bridge Industry are presented in Table 4.4, which presents the top four business strategies for organizational success and the top five characteristics of the competitive environment

According to Manley (2002), these findings and others from the survey support the view that, for the Queensland Road and Bridge Industry, maintaining relationships with existing clients is more important to organizational success than building relations with new clients. The importance of technical capabilities and multi-skilling points to the need for organizations to keep pace with the rapid rate of change and to keep abreast of the increasing knowledge intensity of economic activities. Manley, commenting on the most two most cited characteristics of the competitive environment, concludes that the issue of rapidly changing technologies in the office might reflect the success of government programs to improve the uptake of information and communication technology and that the threat of new competitors reflects the low barriers to entry present in the construction industry. Manley (2002) concludes on the importance of business practice innovations:

> Business practice innovations were highlighted in the survey results, and these sorts of innovations—such as quality systems, human resource strategies and strategic plans—are more people centred that technical solutions to problems. Business practice innovations featured strongly in the identification of 'most important adoptions', and constituted nearly half of the original innovation work. Staff-related strategies were also highlighted as a key tool in maximising the benefits of innovations. These finding highlight the importance of effective 'people policies' to promote higher levels of innovation and improved industry performance. (Manley 2002, p.14)

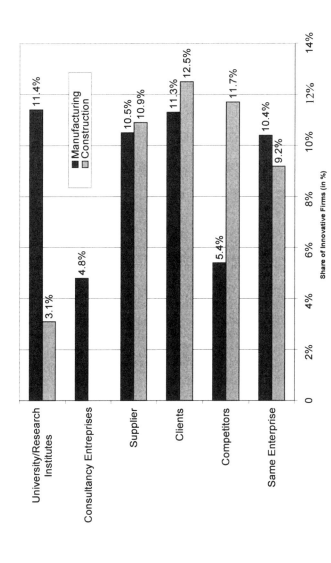

Figure 4.7 Share of co-operation partners of innovators

Source: Cleff and Rudolph-Cleff (2001).

Table 4.4 Top four business strategies for organizational success and the top five characteristics of the competitive environment for the Queensland road and bridge industry

Top-Four Ranked Business Strategies	
Building relationships with existing clients	88%
Delivering products/services that reduce your clients' cost	82%
Enhancing your organization's technical capabilities	80%
Use of multi-skilled teams	77%
Top-Five Ranked Characteristics of the Competitive Environment	
Technologies in the office are changing rapidly	72%
The arrival of new competitors is a constant threat	66%
Our relationships with other organizations in the road industry are assisted by a culture of trust	54%
My clients can easily find a substitute for my services	50%
My organization has many suppliers to choose from	50%

Source: Manley (2002).

4.5 Innovation Determinants

Once a survey has been carried out, construction firms can be divided into two broad categories: Those that are innovative and those that are not - the innovators and the non-innovators. The differences between innovators and non-innovators can be analyzed using econometric techniques which can determine what characteristics of a firm are related to innovative firms and which are related to non-innovative firms. In this manner, one can examine the issue of innovation determinants.

There are two ways to define innovators, depending on whether the survey is an Oslo-type survey or whether the survey is an advanced technology survey. In the case of the Oslo-type survey, innovators are defined as those firms that introduce a new or significantly improved product or process. This section will present as an example a definition of innovators based on an advanced technology and practices survey of construction.

A study by Seaden et al (2001) using data from the 1999 Statistics Canada Innovation, Advanced Technologies and Practices in the Construction and Related Industries analyzed the determinants of innovation in the construction industry. Three measures of innovative activity were developed:

- Innovative technology behaviour (total number of advanced technologies currently used);
- Innovative business behaviour (total number of business practices used);
- Innovative behaviour (total number of advanced technologies and business practices).

The study adopts a measure of innovation that is based on the extent that a firm has adopted advanced technologies and practices. Such a definition of innovation is not consistent with the approach outlined in the Oslo manual. Without presenting the details of the econometric analyses that were carried out, the following results were found:

- Firms that perceived rapid technological change in their environment were more innovative;
- Firms that perceived that materials and supplies quickly became obsolete were more innovative;
- The most innovative of the small and medium-sized firms had strong growth strategies, i.e. market share expansion and geographic expansion were seen as key factors for the success of the firm;
- The most innovative large firms had product range expansion strategies as well as the strategies shared by firms of all sizes (hiring well trained new graduates, developing skills and knowledge of the employees, using multi-skilled teams, improving technology practices/capabilities, and developing proprietary technologies).

The report notes the importance of the culture of firms as a key determinant of innovation. Also of note was the fact that there is a strong association between technology innovativeness and business innovativeness: an innovative firm is generally innovative in technology and in business at the same time: innovativeness tends to be a culture which permeates all the activities of the firm (Seaden et. al. 2001).

4.6. Conclusion

This chapter has looked at the issue of the measurement of innovation in construction. It has pointed out the importance of the definition of construction activities using existing industry classification systems. The results of measurement will be determined by which industries one chooses to measure. Two different ways of defining construction activities were discussed: the construction industry approach and the construction cluster approach.

In terms of innovation, two different approaches were presented: the 'firm as actor' approach and the systems approach. The firm as actor approach focuses on the activities and characteristics of individual firms or other actors. Historically, measurement has focused on the firm as an R&D performer and more recently the focus of attention has been placed on the firm as an innovator. While it is of considerable interest to understanding firm characteristics, the systems approach attempts as well to understand the linkage and interactions between firms and actors.

There are a wide variety of relations that can occur between firms and between firms and other actors, including transactions in the marketplace, and supplier/user relations as well as flows of machinery (embodied knowledge), products, information, manpower and knowledge between actors. Flows are most often not simple transfers but can involve significant engagement and interaction between actors. Collaboration for innovation is a specific case in point. When there are flows, there can be obstacles and these obstacles can be from a wide range of factors, such as lack of user capability or too high levels of risk. The systems approach adds an important dimension to the issue of the measurement of innovation. The two approaches are thus complementary.

Perhaps the most serious gap in the measurement of innovation is the issue of the role of skilled personnel in innovation in construction. Better measure indictors need to be developed in this area. Much of the measurement done to date has concentrated on the actors in the 'construction industry', with some work investigating the relation between manufacturing and the construction industry. The measurement of innovation needs to be broadened to investigate other important actors who have key roles in construction innovation such the service sector, government agencies, building owners and managers, as well as the other actors involved in the various phases in the life cycle of built structures.

Notes

1. One of the major differences between NAICS and NACE (the European) classifications as related to construction is that the NAICS system treats machinery not installed by manufacturers as a construction activity whereas NACE generally treats it as a manufacturing activity
2. The construction sector for the study consisted of all firms with activities in the following NACE-subgroup 45 which include: (451) Site preparation; (452) Building of complete constructions or parts thereof, civil engineering; (453) Building installation; (454) Building completion; and (455) Renting of construction or demolition equipment with operator.
3. The manufacturing sector was grouped into 31 industrial groupings. The industrial groupings discussed in this paper and their corresponding NAICS codes are as follows: Non-Metallic Mineral Products (NAICS 327); Veneer, Plywood and Engineered Wood Products (NAICS 3212); Petroleum and Coal Products (NAICS 324); Electrical Equipment, Appliance and Component (NAICS 335); Other Wood Product Manufacturing (NAICS 3219); Sawmills and Wood Preservation (NAICS 3211); Fabricated Metal Products (NAICS 332); Primary Metals (NAICS 331); Communications Equipment (NAICS 3342); Instrumentation/Magnetic and Optical Media (NAICS 3345 & 3346); Plastics and Rubber Products (NAICS 326); Other

Machinery (NAICS 3333 &3334 & 3335 & 3336 & 3339); Agriculture, Construction and Mining Equipment and Industrial Machinery (NAICS 3331 &3332); Motor Vehicle Manufacturing, Motor Vehicle Body and Trailer Manufacturing, and Motor Vehicle Parts Manufacturing (NAICS 3361 & 3362 & 3363); and Primary Metal Manufacturing (NAICS 331).
4. Industries in NAICS code 23 were surveyed.

References

Note: Statistics Canada publications can be downloaded from http://www.statcan.ca

Anderson, F. (2002), The Flow of Innovative Products from Manufacturing Industries to Construction Industries, Presentation at the *Joint Industry Canada-Statistics Canada Workshop on Innovation Research and Policy Development*, Ottawa, November 1-2, 2002.

Anderson, F. and Manseau, A. (1999), A Statistical Approach to the Measurement of Innovation in Construction, Presentation at the *Third International Conference on Technology Policy and Innovation: Global Knowledge Partnership-creating Value for the 21st Century*, August 30-September 2, 1999, Austin, Texas.

Anderson, F. and Schaan, S. (2001), *Innovation, Advanced Technologies and Practices in the Construction and Related Industries*, Statistics Canada, Science, Innovation and Electronic Information Working Paper, no. 4, February 2001, Catalogue no. 886F0006XIE01004.

Cleff, T. and Rudoloph-Cleff, A. (2001), 'Innovation and Innovation Policy in the German Construction Sector', in André Manseau and George Seaden (eds.), *Innovation in Construction: An International Review of Public Policies*. Taylor and Francis - Spon Press, London.

Dahl, M. and Calum, B. (2001), 'The Construction Cluster in Denmark', in *Innovation Clusters Drivers of National Innovation Systems*, OECD, Paris. pp.179-202.

Den Hertog, P. and Bouwer, E. (2001), 'Innovation in the Dutch Construction Cluster', *Innovation Clusters Drivers of National Innovation Systems*, OECD, Paris. pp. 203-227.

Earl, L. (2002), 'Are We Managing Our Knowledge?: Results from the Pilot Knowledge Management Practices Survey', 2001, SIEID Working Paper no. 6, Cat. No. 880006XIE02006. Statistics Canada, Ottawa.

Eurostat (2000), The Third Community Innovation Survey: Core Questionnaire, 28 November 2000.

Manley, K. (2002), *Innovation in the Queensland Road and Bridge Industry*, Draft Report, commissioned by the Queensland Government, Department of Main Roads, Capacity and Delivery Division, Victoria.

Manseau, A. (1998), 'Who Cares About the Overall Industry Innovativeness?' *Building Research and Information*, vol. 26, no. 4, pp.241-245.

Manseau, A. and Bellamy, A. (1999), 'Canadian Public Policy Instruments that Affect Innovation in Construction', Draft paper, National Research Council of Canada.

McNiven, C. (2002), Use of Biotechnologies in the Canadian Industrial Sector: *Results from the Biotechnology and Development Survey*, Statistics Canada, Science, Innovation and Electronic Information Industry, Working Paper no.3, Cat. No. 880006XIE02006.

OECD/Eurostat (1997), Proposed Guidelines for Collecting and Interpreting Technological Innovation Data (Oslo Manual), OECD, Paris.

OECD (1994), Proposed Standard Practice for Surveys of Research and Development (Frascati Manual), OECD, Paris.

OECD (1999), 'Description of National Innovation Surveys Carried out, or Foreseen, in 1997-99 in the OECD Non-CIS-2 Participants and NESTI Observer Countries', Prepared by Geneviève Muzart, STI Working Papers, Directorate for Science, Technology and Industry, DSTI/DOC (99) 1, 21 May 1999. OECD, Paris.

Peterson, G. (2001), *Electronic Commerce and Technology Use*, Statistics Canada, Science, Innovation and Electronic Information, Connnectiveness Series no. 5, Catalogue no. 57F0004MIE, September, Statistics Canada, Ottawa.

Sabourin, D. and Beckstead D. (1999), *Technology Adoption in Canadian Manufacturing: Survey of Advanced Technology in Canadian Manufacturing, 1998*, Statistics Canada, Science, Innovation and Electronic Information Division, Working Paper no. 5, August 1999, Catalogue no. 88F006XPB, August, Statistics Canada, Ottawa.

Seaden, G., Guolla, M., Doutriaux, J. and Nash J. (2001), *Analysis of the Survey on Innovation, Advanced Technologies and Practices in the Construction and Related Industries, 1999*, Science, Innovation and Electronic Information Division, Statistics Canada, Research Paper Series, Catalogue # 88F0017MIE, No 10, January, Statistics Canada, Ottawa.

Seaden, G. (2001), *Construction Suppliers: Preliminary Analysis of Results from the Statistics Canada 1999 Survey of Innovation*, Report commissioned by the Manufacturing and Processing Technologies Branch, Industry Canada, Ottawa.

Statistics Canada (1996), *Survey of Innovation- 1996*, (Questionnaire). Statistics Canda, Ottawa..

Statistics Canada (1998a), Science and Technology Activities and Impacts: A Framework for a Statistical Information System, Catalogue no. 88-522-XPB, Statistics Canada, Ottawa.

Statistics Canada (1998b), *North American Industry Classification System, 1997,* Catalogue No. 12-501-XPE, Statistics Canada, Ottawa.

Statistics Canada (1999a), Innovation, Advanced Technologies and Practices in the Construction and Related Industries, (Questionnaire), Statistics Canada, Ottawa.

Statistics Canada (1999b), *Survey of Innovation*, (Questionnaire), Statistics Canada, Ottawa.

Vock, P. (2001), An Anatomy of the Swiss Construction Cluster, in OECD, *Innovation Clusters Drivers of National Innovation Systems,* OECD, Paris. pp. 203-227.

Chapter 5

Managing Complex Connective Processes: Innovation Broking[*]

Graham Winch

The aim of this chapter is to provide a review of some important recent contributions to theory and research in the management of innovation with the aim of identifying their relevance for research into innovation in the construction industry. It will become apparent from the review that one of the important issues raised is the role of third parties – organizations that are neither the original source of the innovative idea, nor its implementer – in the innovation process. We have dubbed these organizations 'innovation brokers', and a more detailed discussion of their role follows the literature review before some conclusions are drawn.

The ideas developed in this chapter are drawn from the work of the International Council for Construction Research and Innovation in Building and Construction's (CIB) Task Group 47 (TG47), Innovation Brokering in Construction. The administration of this international task group was funded by the UK's Engineering and Physical Sciences Research Council (award no.GR/R18734/01). Meetings of the TG took place between November 2001 and July 2003 in Manchester, UK; Tokyo, Japan; Cincinnati, USA; Lund, Sweden; and London, UK. The work of TG47 is part of a three phase programme of work, research launched at the London, UK meeting of an earlier CIB task group - TG 35 – in November 1998. TG35 was co-ordinated by André Manseau and George Seaden, culminating in a conference held in Ottawa, Canada in June 2001 and the publication of a source-book on public policies for construction innovation (Manseau and Seaden 2001). Phase Three of the programme of work, focusing on the client's role in innovation, is expected to be launched in Brisbane, Australia in 2004.

---

[*] Funded by EPSRC Award No. GR/R18734/01

5.1 The Conventional Model of Innovation: A Sketch

For over 20 years, innovation research has been dominated by the model developed by Abernathy and Utterback (well summarized and developed in Utterback 1994). The central tenet of this model is that of the industry life-cycle from an early fluid phase when product enhancing innovation is the key, and where there is intensive competition between different design concepts, through a transitional phase which witnesses the emergence of a dominant design to a specific phase when competition is between a few large firms through performance improving (particularly cost reducing) innovations. While there has been much debate over the details of this model, and the differences, for instance, between assembled and non-assembled products, the basic life-cycle model is widely accepted, and can currently be seen at work with a vengeance in the market for internet service providers.

An implicit assumption of this model is that of the single firm – or population of competing firms – striving for the attentions of the customer through the market (Hobday 1998). Through this process of creative destruction, 'the ecology of competing firms changes from one characterized by many firms and many unique designs, to one of few firms with similar product designs' (Utterback 1994 p 24). There is a tradition from Schumpeter onwards of castigating the construction industry – and particularly housebuilding - for not following this model and becoming the 'one industry that God forgot and the industrial revolution overlooked' (cited Lawrence and Dyer 1983 p 158). However, recent research has suggested that there is large number of industrial sectors where this model has little analytic leverage. Where industries do not produce discrete products, but 'large technical systems' (Coutard 1999), or 'complex systems' (Miller *et al* 1995), then a rather different approach to innovation is required. Even in the case of housing, this standard model appears to be inadequate. This chapter aims to review some of these recent contributions and to identify their relevance for those concerned with construction innovation. In particular, the role of innovation brokers in innovation will be identified, and their contribution to innovation processes in construction assessed.

5.2 Emerging Perspectives on the Management of Innovation

This section briefly reviews a variety of different perspectives, and evaluates their contribution from the perspective of the role of innovation brokers in construction innovation. These are not intended to be comprehensive reviews, rather attempts to make connections between work in progress by leading innovation research groups and the concerns of researchers focused on innovation in the construction industry.

Complex Products and Systems (CoPS)

Pioneering research during the early 1990s by a team from l'Université de Québec à Montréal and the University of Sussex (Miller *et al* 1995) developed the notion of complex product systems and the complex systems industries that produce them. In the UK, this work has been further developed in a number of ESRC and EPSRC funded projects at the Science Policy Research Unit, University of Sussex. The approach is case-study based, and covers a number of complex products, ranging from high speed trains to turbo-prop engines. A special issue of *Research Policy* vol. 29 (2000) reports much of this research, while Davies (2001) provides an overview of some more recent work. There are a number of themes of importance to those concerned with innovation in construction, of which the following are, perhaps, most relevant:

- Technologies evolve over time, and have to be negotiated both amongst the suppliers of the CoPS, and also with clients and regulators. This evolutionary characteristic means that there is rarely radical shake-out in the CoPS industries due to technological shifts. The core competences of CoPS suppliers are the combination of project management capabilities and systems integration capabilities;
- Many CoPS form parts of larger technical systems such as electricity supply networks, which impose further constraints on the innovation process;
- Increasingly, CoPS suppliers no longer only supply the hardware (capital goods), but also the services associated with the exploitation of that hardware, and clients shift to buying system availability, rather than the system itself. The income stream from the service element can be a multiple of ten or more of that from the supply of the product itself;
- IT is an increasingly integral part of all CoPS, not only those that are themselves IT systems;
- CoPS suppliers are paying increasing attention to the management of portfolios of supply projects and ensuring learning from them, rather than managing projects as one-off entities. Learning from projects is the key management challenge for CoPS supplier firms.

Co-operative Technical Organizations

Work very much complementary to the Miller and Hobday work on flight simulation was undertaken by Rosenkopf and Tushman (1994; 1998). They focused specifically on the role of co-operative technical organizations (CTOs) in innovation in the flight simulation industry. These are the collaborative organizations which bind together the firms in an innovation network. They may be temporary or permanent in nature, and vary significantly in constitution, but they all serve to facilitate innovation where firms need to agree upon standards and interfaces in order to have the confidence to commit to the development of innovative ideas.

Taking their analytic perspective from organizational ecology (Aldrich 1978) Rosenkopf and Tushman analyze the evolution of CTOs between 1958 and 1992 using network analysis techniques. They identify a large number of different types of CTOs, which can be grouped into cliques of overlapping membership. These include professional institutions, standards bodies, regulators, trade associations, and various ad hoc bodies. They also report that the foundation rate of CTOs was much greater during periods of radical technical change and slower during periods of incremental change.

Distributed Innovation Processes (DIPS)

It has long been recognized that innovation by a single actor (firm) is very rare – innovations are the result of collaboration between firms in clusters or networks. Examples of this include the importance of universities and government laboratories for sources of new ideas which are then exploited commercially by firms, the role of demanding customers in stimulating supplier firms, and various forms of inter-firm collaboration particularly in supply chains. The literature on national systems of innovation can be seen as taking this perspective and applying it at the institutional level, rather than the firm level. These collaborative ventures between firms can be defined as DIPS (Coombs *et al* 2001) and within this perspective the innovation processes associated with CoPS are special cases of DIPS.

However, this established view of DIPS is inherently static – it only explains what exists, not how DIPS evolve and decline. It shares this weakness with another important influence in innovation research – Schumpeter's analysis of innovation and competition, which sees the process of creative destruction as product markets moving from one stable state to another after the introduction of a radical technology. However, the period of turmoil between the two stable states is rarely analyzed, and the sources of radical innovation remain external to the analysis.

Work at the Centre for Research on Innovation and Competition at UMIST is attempting to build on these two existing bodies of work by developing a dynamic analysis of DIPS. A stable DIP consists of a network of firms embedded in a particular social context, where innovations by the various participants are interdependent, and innovations yield returns for the innovator, but greater returns for the firms in the DIP in aggregate. However, if one member innovates in such a way that its appropriable returns are greater than the aggregate, then the DIP becomes unstable and, over time shifts to a new configuration. Competition can also exist between DIPS, where some configurations may be more effective at delivering innovations to market than others.

Games of Innovation

Work at the Ecole Polytechnique, Montréal, in collaboration with the Industrial Research Institute, is investigating how different *games of innovation* generate competitive advantage for firms through their research and development (R&D) activities (Miller and Floricel *forthcoming*). The research is based on a pilot stage of interviews with 70 Chief Technology Officers (CTOs) or Vice-Presidents for R&D, followed by a worldwide survey of 1000 firms – the research is currently in the latter phase.

Games of innovation are the distinctive configurations of innovation practice in particular industries and markets driven by their predominant ways of creating value in interaction with their social, technical, economic and political (STEP) environment. Preliminary factor analysis of the data has identified 8 generic games, grouped into high and low velocity games:

High velocity

- *Battles of architectures*, such as consumer software and electronics products;
- *Delivering safe science-based products*, such as in pharmaceuticals
- *Information systems design and consulting services*;
- *Races to the patent and regulatory offices*, such as in biotechnology;
- *Research, development and engineering services*, where firms carry out commissioned R&D such as designing new computer chips.

Low velocity

- *Delivering workable solutions in packs*, such as building materials suppliers;
- *Asset-specific problem-solving*, such as operators of utilities and petrochemical refiners;
- *Customized high-tech craft*, such as aerospace.

There is general agreement[1] that construction is more in the low-velocity category, with its emphasis on problem-solving for clients, and its emphasis on engineering design. Perhaps the following observations can be inferred from these early results of the research programme:

- Asset-specific problem solving firms form a significant part of the client-base for construction and engineering construction firms;
- Customized high-tech craft firms are those which fit the complex systems industry model most closely;
- Asset-specific problem-solving would appear to relate to the processes that Hughes (1983) identified in his analysis of power systems.

The analysis of the differences between high-velocity and low-velocity games is sophisticated, and not yet complete. However, some of the emerging themes which particularly characterize low-velocity games relevant to our concerns are:

- The relative importance of 'roadmap' research, compared to the search for new technologies, particularly for customized high-tech craft and asset-specific problem-solving firms;
- The relative importance of clients as dominant stakeholders in the innovation process for customized high-tech craft firms;

- The comparative importance of 'managing integration processes', 'social regulation', corporate governance, and project management capabilities as levers of innovation, while science-based R&D is relatively unimportant;
- The relative unimportance of universities as the sources of new ideas;
- The importance of working in networks and through industry technical organizations such as standards bodies, particularly for asset-specific problem-solving firms.

The work does not address the role of innovation brokers as such – it is focused on the innovating firm. However, there is some analysis of the role of networks in innovation, and the ability to integrate across inter-organizational networks appears to be particularly important in low-velocity innovation, while collaborating with partners and clients also appears to be relatively important. For those concerned with construction innovation, the strength of this work is its comparative nature, allowing us to identify more easily how industries similar to construction differ in their innovation games from the technologically dynamic sectors characterized by high-velocity innovation games.

Regional Clusters and Social Capital

Research conducted by the Centre for International Studies at the University of Toronto draws our attention to the geographical aspects of innovation. A considerable amount of research on economic performance generally (e.g. Porter 1990), and innovation in particular has investigated the ways in which successful firms cluster geographically. This strand of research has also been linked to the concept of national business systems (e.g. Whitley 1999), and national systems of innovation (e.g. Lundvall 1993), and has more recently been applied at the more local level in the analysis of regional systems of innovation (Wolfe 2002).

There are a number of differing approaches to these issues, but they share a common basis in emphasising the context of economic activity, and the way it is embedded (Granovetter 1985) in a social and institutional context. In this view, firms are part of geographically specific networks that nurture and facilitate both their current competitiveness in international markets and their ability to innovate to ensure that their competitiveness is sustainable. Arguably, as globalization develops, regional specializations, in particular goods and services become even more important.

Research for the European Union has identified a number of important elements of such regional innovation systems, including:

- Technology centres;
- Technology brokers;
- Business innovation centres;
- Research universities;
- Venture capitalists.

These are the sorts of organizations that governments can sponsor, and which provide immediate innovation-related services to firms in the region. Other researchers have placed greater emphasis upon more underlying features of the regional innovation system – in particular the institutions and cultures that enable the development of social capital and trust between firms. *Social capital* is the web of interrelationships that allows business to be done on the basis of trust, rather than contract, thereby generating efficiencies by allowing considerable savings in transaction costs. These relationships are generated through the development of community organizations – business orientated or not – ranging from chambers of commerce to golf clubs; shared educational experiences; and, some would argue (e.g. Fukuyama 1995), shared cultural values.

Technology Fusion

Research in Japan on the R&D strategies of leading Japanese corporations has identified the importance of a *technology fusion* approach to R&D, as opposed to a *breakthrough* strategy (Kodama 1992). Under the breakthrough approach, firms try to make the next leap forward in technology either through their own efforts or in industrially orientated collaborative associations – the defence-driven US high technology programmes are cited as the prime example of the breakthrough approach. Successful Japanese firms, on the other hand, now focus on making smaller incremental technological advances which are turned into radically new products by bringing together advances from various different technologies such as the development of mechatronics (e.g. stepping motors for automated control systems) and optoelectronics (e.g. fibre-optic cables). The key to technology fusion is collaborative R&D with firms from *outside* the market sector concerned, led by broad search and strong in-house engineering capabilities. Industry-level R&D research associations are becoming less important as leading firms range more widely in their search for new technologies.

Implications for Construction Innovation Research

The importance of innovation brokers was first identified with compelling clarity in the work on the flight simulation industry. Miller and his colleagues (1995) identified the importance of professional institutions – in this case the UK's Royal Aeronautical Society in providing an international neutral space for broking innovations. Their findings were greatly reinforced by the analysis of Rosenkopf and Tushman (1998) on cooperative technical organizations. These concepts have been applied to the construction industry (Winch 1998), and Hobday (1998) also argues that a number of constructed products are CoPS. In essence, the proposition is that the process of innovation in construction cannot be understood by borrowing analytic frameworks from the industries to which the life-cycle model applies, but by developing an analytic framework appropriate to the needs of construction as a complex systems industry. In the past, construction has been castigated as 'backward' simply because of its failure to make the transition from fluid to specific. The proposition here is that it shares characteristics with a number of other capital goods sectors – particularly high-tech ones – which mean that it is qualitatively different, not backward. This model is illustrated in Figure 5.1.

There are three main elements to construction as a complex systems industry, the *innovation superstructure*, which is, effectively, the 'demand side' for innovation, while the *innovation infrastructure* represents the 'supply side', and is typically the source of new ideas.[2] These two elements are linked through the *systems integrator* role, which manages the interaction between market demand and technological possibility. The superstructure and infrastructure each consist of a number of different types of organization as follows:

Innovation superstructure

- A *client*, without which production does not start;
- *Regulators*, which ensure that the interests of external stakeholders, such as public in the structural integrity of a building, are expressed;
- *Innovation brokers* that facilitate innovative activity by the innovation infrastructure - trade contractors, systems integrators, specialist consultants, and component suppliers.

Innovation infrastructure

- *Trade contractors* which supply and fix specialist sub-systems;
- *Specialist consultants* which design specialist sub-systems;
- *Component suppliers* which supply components specified by consultant designers and fixed by trade contractors that make up specialist sub-systems.

Systems integrators

- These articulate client requirements in relation to technical possibilities. Systems integrators both design the overall product concept and engineer the interfaces between technical sub-systems, and project manage delivery of the product to the client.

A comparison of this figure with the innovation structure in flight simulation (Miller *et al* 1995 Fig 2) reveals an important feature of the construction industry. In construction the project management and systems integration capabilities are typically shared between the contractor on the one hand, and the consultant designers on the other. In contrast, most other CoPS supply sectors integrate the systems integration and project management capabilities required in single firms that compete for turnkey projects. The implications of this split for innovation processes in construction warrant further investigation.

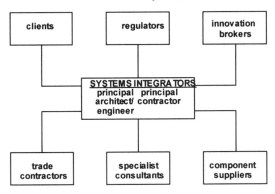

Figure 5.1 Construction as a complex systems industry

Source: Adapted from Winch (1998 Figure 1).

However, it can be argued that not all constructed products are CoPS. Within Hobday's (1998 table 1) illustrative list of CoPS are: bridges; clean rooms; dams; docks and harbours; intelligent buildings; intelligent warehouses; nuclear power plants; nuclear waste storage facilities; road systems/flyovers; runways for aircraft; sewage treatment plants; rail transit systems; and water supply systems. These are all highly engineered constructed products, typically deploying leading edge technology. Simpler and more routine types of constructed products, ranging from schools to housing are, arguably, not CoPS. This is the importance of the DIPS approach.

Although the DIPS analysis does not explicitly identify or analyze the role of innovation brokers, it would appear to be implicit in the attention paid to inter-firm relations, and the social embededness of DIPS in networks of 'concertation' between firms. The perspective offers the advantage that it could encompass the whole of construction, including housing, with CoPS as a special case. A network perspective on innovation in construction underlies the work of Miozzo and her colleagues (e.g. Miozzo and Dewick 2002).

Clearly, the site specific nature of construction means that a geographical approach to the role of innovation brokers is a vital part of the analysis; particularly those innovation brokers aimed more at housing and building firms and construction SMEs in general, rather than major projects where firms tend to work more in national and international markets. An interesting aspect of this is the way in which some universities provide support for the foundation and development of SMEs by providing income from part-time teaching – the tendency of young design firms to cluster around architecture schools is an example of this[3] (see Williamson 1990). Turning to the larger firms, a further interesting question is whether there are geographic clusters of construction firms. Evidence from ENR data suggests that there may be, for at least those firms in the market for internationally traded construction services associated with major projects (Rimmer 1988). A contemporary re-analysis of the data might suggest that London, Tokyo, Paris, and the Randstaat region of the Netherlands act as clusters for the major projects sector[4], which raises the question of whether any cluster-specific innovation brokers have emerged.

It can be argued that the technology fusion approach is much more appropriate for construction R&D than the more widely accepted break-through approach (Groák 1994). This is because it is not very appropriate analytically to conceptualize construction as 'an industry', when in fact it is an aggregation of projects aimed at meeting client needs by drawing on appropriate technologies whatever their provenance. While this is certainly true of the major projects sector, it is arguably less true of the building and housing sectors[5], while the repair and maintenance sector, with its inherent bias towards historic technologies, has its own special needs. Tomonari Yashiro has suggested[6] that while the systems integrator is best placed to achieve technology fusion, the innovation broker can provide a source of inspiration for 'fusion-makers'.

5.3 The Concept of an Innovation Broker

An important issue is the relationship of the concept of an 'innovation broker' to that of 'knowledge broker' which is also gaining currency. There are perhaps three positions on this:

- Knowledge brokers are a for-profit sub-set of innovation brokers, which include many not-for-profit organizations. For instance, engineering consultancies that capture and diffuse innovative ideas between firms have been defined as 'knowledge brokers' (Hargadon 1998). However, it might be suggested that such firms are more important in their roles as specialist consultants, and only rarely play a true broking role;
- Knowledge brokers are a much larger group of organizations concerned with the diffusion of knowledge, of which innovation brokers are a sub-set (Bang *et al* 2002). Such organizations include universities in their teaching, as opposed to research roles. However, such organizations contribute little, as such, to innovation. Although highly educated staff are a necessary condition for an adequate absorptive capacity, it is not a sufficient one; the appropriate organizational processes must also be in place;
- Conceptually, there is little difference between the two. Knowledge is a fundamental prerequisite for innovation, and, in any case, many of the

activities of innovation brokers as defined within the CoPS framework are learning from projects and the diffusion of those lessons to other firms within the sector.

Discussions[7] within TG47 have concluded that an organization – whether for profit or not - needs to meet the following criteria to be categorized as an innovation broker:

- It acts as an intermediary between sources of innovative ideas and implementers, including the situation where potential implementers were also potential sources;
- It provides authoritative validation of information. Arguably, it is through this validation process that brokers potentially add most value;
- It is trusted by implementers as being objective in this validation;
- It helps innovators to diffuse research results, but pure diffusion is not a brokerage activity.

Based on these considerations, the TG47 operational definition of 'innovation broker' can be formalized thus:

> An organization acting as a member of a network of firms in an industry that is focused neither on the generation nor the implementation of innovations, but on enabling other organizations to innovate.

Many organizations in such networks will have more than one role. Some brokers also act as a generator of innovations; less frequently they may act as an implementer.

5.4 An Innovation Broker's Research Agenda

A central part in developing this framework is identifying the type and modalities of the innovation brokers in construction as they form one of the main elements of the innovation superstructure of the national construction innovation system. Influenced by work on the flight simulation industry, as well as earlier work on construction innovation, a first-cut analysis would include:

- National building research institutes;
- Professional associations;
- University institutes;
- Trade associations;
- Standards organizations;
- State-sponsored innovation bodies.

George Seaden[8] has suggested one classification of knowledge brokers into:

- Those independent centres employing staff with a large range of research skills, and well-found experimental and testing laboratories (e.g. Building Research Establishment, or Centre Scientifique et Technique du Bâtiment offering a full range of services to industry);
- Those connected with research universities (e.g. Center for Integrated Facilities Engineering, Stanford University) which offer much more specialist research capabilities at the leading edge of scientific advance;
- Special-purpose applied technology institutes, such as the Collaborative Research Centre Construction in Brisbane;
- Networked technology advisory services, which tend to be aimed more at SMEs and staffed by people with extensive industrial experience who work more in a consultancy mode such as the Centre for Construction Innovation in Manchester.

Gann (2001) discusses the potential for professional institutions, such as the UK's Institution of Civil Engineers, to play the role of innovation broker. In other sectors, such as flight simulation, they do play an important role. However, in construction there are, arguably, too many of them competing for influence, and they have a tendency to be insular and defensive in an industry undergoing rapid change. Typically, they lack the capacity to capture and exploit new ideas, and find it difficult to attract people with appropriate skills. However, they do have potential to act as repositories of knowledge in a project-based world if they were to return to their 19th century roots as learned societies, particularly if they can bring a more international perspective to their role.

As this discussion suggests, not all construction innovation brokers are the same in both aims and operation. It might be suggested that there are various dimensions that we will need to consider in our research:

- The national construction business system, and system of innovation;
- Mode of innovation addressed;
- Position in the innovation diffusion process.

5.5 National Construction Business System

It might be expected that the relative role and importance of different types of innovation broker would vary between countries. For instance, in Anglo-Saxon type countries, where autonomous professional associations are relatively important, it might be expected that these would play a larger role than state organizations. The reverse might be found in *étatique* type countries. Certainly, Swan and her colleagues (1999) found that professional associations played different roles in the diffusion of CAPM systems in the UK and Sweden, and that this was related to the different role of the state in innovation brokering. Similarly, Hughes (1983) shows how independent consultants (Stone & Webster in the US and Merz & McLennan in the UK) played key roles in the development of another CoPS – electricity supply systems – for their countries, while state (*Land*) agencies played similar roles in Germany.

The type of innovation activity favoured by firms may also be linked to the type of business system. For instance, Miozzo and Dewick (2002; *forthcoming*) suggest the Anglo-Saxon characteristics of market orientated short-termism displayed by UK construction firms encourages them to focus on process-orientated research, enabling them to develop their project management capabilities and to focus on the relationships between the parties. Those firms that benefit from more 'patient' capital such as the French, German and Swedish tend to favour longer-term programmes of R&D focused on developing their operational capabilities on site. This research tends to focus on the construction and performance of concrete structures and IT applications and they also tend to make significant efforts at knowledge management within the firm. Where cross-membership of boards between clients and the supply chain is common, such as in Sweden and Germany, then supply chain innovations tend to be facilitated. The much smaller size of Danish firms appears to create a situation more similar to the UK, where the relationships between the parties are emphasized. Interestingly, it is in Denmark and the UK that relationships with the government are strongest in the area of R&D.

5.6 Mode of Innovation Addressed

In a review of the management of innovation in construction, Winch (1998) identified two main modes of construction innovation as illustrated in Figure 5.2. The top-down mode is where the management of innovation issues revolve around the adoption of new ideas by firms, and their implementation on projects. The bottom-up mode is where the management of innovation issues revolve around problem-solving on projects, and the learning of those solutions by the firm for implementation on further projects.

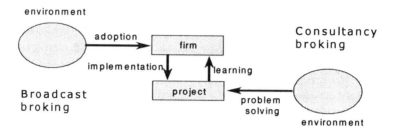

Figure 5.2 **The top-down and bottom-up modes of construction innovation**

Source: Developed from Winch (1998).

The role of innovation in the adoption mode is likely to be much more proactive, and to use promotional techniques. The principal aim of such an innovation broker is to promote awareness, and facilitate the implementation of new ideas on projects. Such innovation brokers, it can be suggested, are more likely to work in a *broadcast* mode. Problem-solving on projects, on the other hand, is likely to require a very different kind of innovation broking. Brokers will need to be more reactive, but backed with a depth of technical expertise to find the best solution for innovators. Here innovation broking is likely to be much more in a *consultancy* mode.

5.7 Position in the Innovation-diffusion Process

Figure 5.3 presents a model of the diffusion of *practices* in construction, Here, the concept of a 'practice' is analogous to Nelson and Winter's concept of an operational routine, where a routine[9] is a 'regular and predictable behaviour pattern' (1982 p 14). This definition connects with the concept of 'best practice', which is widely, if loosely, used by innovation brokers and their sponsors. Best practice is only meaningful as a point on a distribution of practices in the manner shown in Figure 5.4.

Developing this definition, it might be suggested that ahead of best practice is *advanced practice*, which has experimental elements, and may not be fully proven – firms that are effective at advanced practices may possess significant competitive advantages, because, by definition, advanced practice is distinguishable from the competition. *Best practice* is the proven suit of practices that have diffused to a proportion of the most competitive firms; they are typically the focus of the most active promotion by brokers concerned with top-down innovation. Early movers with best practices find that they do possess competitive advantage, but in its very nature, competitive advantage derived from best practices is not sustainable as they become more widely adopted. *Standard practices* are market-entry criteria – they summarize the minimum standards that must be met, and they are typically enshrined in codes and regulations. Firms using unacceptable practices – be they corrupt or incompetent – do not possess competitive advantages, and are obliged to compete on price, and tend to work for only the more undiscerning clients[10].

In this model, practices are dynamic – yesterday's best practice is today's standard practice. This proposition is central to evolutionary economics where and firms which fail to adopt the routines appropriate for their market slowly die, or in industry such as construction, where exit barriers are high[11], slowly decline into unacceptable practice. Arguably, what distinguishes 'best practice' firms from 'standard practice' firms over time is their greater absorptive capacity – their greater ability to learn and implement new ideas. Advanced practice firms, it can be argued, need to posses the greatest levels of absorptive capacity, for they are likely to be working on the most challenging projects for clients, and to be ahead of the game in adopting and implementing new practices.

Figure 5.3, is, in effect, an innovation/diffusion model as new ideas are managed into good currency. However, this is not a linear model of moving from new ideas on the left of the diagram to full diffusion on the right. Innovation can occur at any point in the diffusion process, as reinvention (Rice and Rogers 1980) takes place. What changes from left to right is the level of uncertainty associated with the innovation. Thus the process of achieving good currency has been broken into five phases, which are conceptual rather than linear, and in reality have much more interaction between them than can be shown in a single diagram. New ideas can be generated at any of the first four stages, but as the process moves from left to right in the diagram, the level of codification increases, and the innovations become more and more incremental in character.

96   *Building Tomorrow*

Figure 5.3   The diffusion of practice in construction

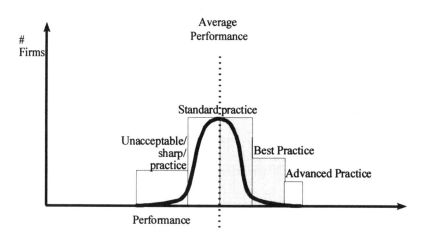

Figure 5.4   Practices and the diffusion of innovations

Building on the model presented in Figure 5.4, Widén suggests[12] that the purposes of innovation brokers differ, depending on whether they are concerned with the early phases of the diffusion process, or the later ones. In the earlier phases they are less proactive, acting very much as a facilitator of the process by providing neutral spaces for innovators to thrash out their ideas. Later in the process, brokers tend to be more proactive, validating and promoting innovations to later adopters of best and standard practices. Value is added by both types of broker, but it is more obvious for the latter type. These points can be linked to the work of Hansen and his colleagues (1999) on knowledge management. They argue that where markets for services are diffused and require bespoke treatment, a *personalization* approach to knowledge management is more appropriate. Where they are amenable to standardization of the service provided, then *codification* is a more appropriate approach. Personalization requires intensive inter-personal interaction, with IT as a background enabler. Codification, on the other hand, requires extensive use of IT to capture and diffuse learning within the organization.

5.8 Concluding Thoughts

This chapter has reviewed some of the latest research in innovation which challenges the dominant model of innovation derived from work in the volume consumer products sector. Taken as a whole, this research suggests that there are other models of successful innovation which can be used to analyze the construction case with greater acuity than the dominant model. A theme running through much of this recent research is that innovation takes place within networks of competing and collaborating firms, rather than within a single firm. Within such networks, the role of innovation brokers as third parties in the diffusion and process is receiving growing attention.

Drawing on this work, the chapter has provided a definition of 'innovation broker' as applied to the construction case, and suggested some of the research issues around such organizations. It should, however, be stressed that we are not claiming that innovation cannot take place without brokers. Many large firms have the capability to go directly to the sources of new ideas in universities and elsewhere, or to generate new ideas themselves, particularly firms playing high velocity games of innovation. However, this approach would appear to be relatively rare in construction compared to other industries, perhaps due to the lack of dominant firms in the industry, and the relatively low levels of absorptive capacity (Cohen and Levinthal (1990) in construction firms (Gann 2001). Arguably, innovation brokers are more important elements of the innovation process – and hence of the national innovation system – in construction than in other sectors.

Notes

1. Discussion at the TG 47 Manchester seminar, November 2001.
2. Of course, these organizations may in turn, obtain new ideas from other sources such as research institutes and universities.
3. For instance, schools such as the Architectural Association and The Bartlett in London have played an important role in launching the practices of many of today's most innovative architects such as Richard Rogers and Norman Foster by providing teaching opportunities.
4. US firms dominate the international market for construction services, but they are regionally dispersed within the US, and a similar argument applies to Germany. The UK cluster would include foreign-owned firms such as Bechtel and Skanska which have major international operations based within the London region. The clusters specified would probably be reinforced if project finance companies were also included in the major projects cluster.
5. See Winch (2003) for a rationale of this sectoral breakdown.
6. Presentation at the TG 47 Tokyo meeting, February 2002.
7. At the TG 47 Lund meeting, February 2003.
8. Presentation at the TG 47 Manchester seminar, November 2001.
9. The term routine is not retained, because Nelson and Winter suggest that a 'routine' is the economic counterpart of the biological gene – no such analogy in intended here.
10. It might be argued that unacceptable practice has no part in a model of the diffusion of innovation, as such firms are unlikely to be adopting new ideas. There is much to be said for this argument, but if the concern in firm performance as an outcome of innovation, rather than innovation per se, then it does have a role. From this perspective clients can get the quickest improvement in supplier performance simply by avoiding firms using unacceptable practices. Arguably, policies aimed at 'chopping off the tail' of poorly performing firms can have a quicker impact on aggregate firm performance than more positive policies of promoting innovation.
11. This statement is, at face value, improbable. However, it is based on the observation that firms continue to survive in construction at the margins of the industry in a way that that they would not in many other sectors. This suggests high exit barriers such as lack of transferability of skills and low profitability which makes it difficult to buy into more promising sectors – in other words, construction firms easily get caught in to a vicious circle of decline into 'sole practice' as a one-person architectural practice or labour-only sub-contractor.
12. Discussion at the TG 47 Cincinnati meeting, September 2002.

References

Aldrich, H. (1978), *Organizations and Environments,* Englewood Cliffs, Prentice-Hall.
Bang, H. Clausen, L. and Dræbye, T. (2002), 'Mapping of Knowledge Brokers: The Case of Danish Building', In: *Construction and Global Competitiveness: Proceedings 10th International Symposium: The Organization and Management of Construction,* CRC Press, Boca Raton Florida.
Cohen, W. M. and Levinthal, D. A. (1990), 'Absorptive Capacity: A New Perspective on Learning and Innovation', *Administrative Science Quarterly,* vol. 35.

Coombs, R., Harvey, M., and Tether, B. (2001), *Analysing Distributed Innovation Processes*, Manchester, Centre for Research on Innovation and Competition.
Coutard, O. (ed.) (1999), *The Governance of Large Technical Systems*, Routledge, London.
Davies, A. (2001), *Integrated Solutions: The New Economy Between Manufacturing and Services*, Science Policy Research Unit, University of Sussex, Brighton.
Fukuyama, F. (1995), *Trust: The Social Virtues and the Creation of Prosperity*, Hamish Hamilton, London.
Gann, D. (2001), 'Putting Academic Ideas into Practice: Technological Progress and the Absorptive Capacity of Construction Organizations', *Construction Management and Economics*, vol. 19, pp. 321-330.
Granovetter, M. S. (1985), 'Economic Action and Social Structure: the Problem of Embeddedness', *American Journal of Sociology*, vol. 91, pp. 481-510.
Groák, S. (1994), 'Is Construction an Industry? Notes towards a Greater Analytic Emphasis upon External Linkages', *Construction Management and Economics*, vol. 12, pp. 287-293.
Hobday, M. (1998), 'Product Complexity, Innovation and Industrial Organization', *Research Policy*, vol. 26, pp. 689-710.
Hughes, T. P. (1983), *Networks of Power: Electrification in Western Society 1880-1930*, Johns Hopkins University Press, Baltimore.
Kodama, F. (1992), 'Technology Fusion and the New R&D', *Harvard Business Review*, vol. 70, no. 4, pp. 70-78.
Lawrence. P. R. and Dyer, D. (eds.). (1983), *Renewing American Industry*, Free Press, New York.
Manseau, A. and Seaden, G. (eds.). (n.d), *Innovation in Construction: An International Review of Public Policies*, Spon, London.
Miller, R. and Floricel, S. (forthcoming), 'Value Creation and Innovation Games', *Research Technology Management*.
Miller, R, Hobday, M., Leroux-Demers, T. and Olleros, X. (1995), 'Innovation in Complex Systems Industries: the Case of Flight Simulation, *Industrial and Corporate Change*, vol. 4, pp. 363-400.
Miozzo, M. and Dewick, P. (2002), 'Building Competitive Advantage: Innovation and Corporate Governance in European Construction', *Research Policy*, vol. 31, pp. 989-1008.
Miozzo, M. and Dewick, P (2004), 'Networks and Innovation in European Construction: Benefits from Interorganizational Cooperation in a Fragmented Industry', *International Journal of Technology Management* vol. 27, no. 1, pp. 68-93.
Nelson, R. R. and Winter, S. G. (1982), *An Evolutionary Theory of Economic Change*, Cambridge, Belknap.
Rice, R.E. and Rogers, E.M. (1980). 'Reinvention In The Innovation Process' *Knowledge-Creation Diffusion Utilization* Vol. 1, pp. 499-514.
Rimmer, P. J. (1988), 'The Internationalization of Engineering Consultancies: The Problems of Breaking into the Club', *Environment and Planning*, Vol. A 20.
Rosenkopf, L. and Tushman, M. L. (1994), *Community Organization and Technological Evolution: Inter-organizational Cooperation over the Technology Cycle*, Wharton School, University of Pennsylvania.
Rosenkopf, L. and Tushman, M. L. (1998), 'The Co-evolution of Community Networks and Technology: Lessons from the Flight Simulation Industry', *Industrial and Corporate Change*, vol. 7, pp. 311-346.

Swan, J., Newell, S. and Robertson, M. (1999), 'Central Agencies in the Diffusion and Design of a Technology: A Comparison of the UK and Sweden', *Organization Studies,* vol. 20, pp. 905-932.

Utterback, J. M. (1994), *Mastering the Dynamics of Innovation,* Boston, Harvard Business School Press.

Van de Ven, A. H. (1986), 'Central Problems in the Management of Innovation', *Management Science,* vol. 32, pp. 570-607.

Williamson, R. (1990), *American Architects and the Mechanics of Fame,* Austin, University of Texas Press.

Winch, G. M. (1998), 'Zephyrs of Creative Destruction: Understanding the Management of Innovation in Construction', *Building Research and Information,* vol. 26, pp. 268-279.

Winch, G. M. (2003), 'Models of Manufacturing and the Construction Process: The Genesis of Re-engineering Construction', *Building Research and Information,* vol. 31, pp. 107-118.

Wolfe, D. (2002), 'Social Capital and Cluster Development in Learning Regions In: J. Adam Holbrook and David A. Wolfe (eds), *Knowledge, Clusters and Regional Innovation,* Kingston: Queen's University School of Policy Studies, University of Toronto.

Chapter 6

Understanding Risks and Shaping Large Engineering Projects

Roger Miller and Serghei Floricel

Large engineering projects (LEPs) are high-stakes games characterized by substantial irreversible commitments, skewed reward structures, and high probabilities of failure. Once built, projects have little use beyond the original intended purpose. The journey to the period of ramp-up and revenue generation takes ten years on average. Substantial front-end expenditures prior to committing large capital costs have to be carried by sponsors; these expenditures may sum up to 33 per cent of the total cost for some projects. But despite these huge investments in planning, the reality of market demand and the true worth of the project can be truly tested only during the ramp-up period. Sponsors may find that reality is very different from expectations.

Managing risks is thus a real issue. The purpose of this chapter is to sketch-out the various components of risk and outline the strategies for coping with risks and turbulence. The main thesis is that successful projects are not selected but shaped. Rather than evaluating projects at the outset based on projections of the full sets of benefits, costs and risks over their lifetime, successful sponsors start with project ideas that have the possibility of becoming viable. Then, they embark on shaping efforts to influence risk drivers from project-related issues to broader institutional elements. The seeds of success or failure of individual projects are thus planted early and nurtured over the course of the shaping period as choices are made. Successful sponsors however do not escalate and abandon quickly when they recognize that projects have little possibility of becoming viable.

6.1 The IMEC Study and Large Engineering Projects

The IMEC's (International Program in the Management of Engineering and Construction) study grew out of the noted difficulties that project delivery encountered in meeting targets. As long as demand for infrastructure pulled supply in a manageable way, governments and businesses were content to rely on traditional management methods. However, as demand for better risk analysis increased and public financing became tight, methods that had served their purpose in the past were no longer sufficient.

IMEC was thus set up to understand the changes that were occurring (Miller and Lessard 2001). At the time (mid-1990s), to our knowledge, there had been no attempt to study, evaluate, and present a systematic analysis of the new approaches to large projects. Hence, we decided to undertake grounded research to understand what leads to success or failure, using a sample of 60 LEPs. The goal was to identify the practices that, in the experience of executives involved in projects, really made a difference. Here are the distinctive features of the IMEC research program.

An International Field Study

The study sums up exclusive collective experience. In general, seven to eight players—sponsors, bankers, contractors, regulators, lawyers, analysts, and others—were interviewed for each of the 60 projects.

Systemic and Strategic Perspectives

Particular emphasis was placed on front-end development decisions, but execution and initial ramp-up top operation were also studied. Calling upon a range of disciplines, the IMEC study focused on themes such as coping with uncertainty through risk analysis, institution shaping, and strategies.

Diversified Range of Domains

The 60 projects included fifteen hydroelectric dams, seventeen thermal and nuclear power plants, six urban transport facilities, ten civil infrastructure investments, four oil platforms and eight technology initiatives.

6.2 Projects Differ Substantially in Terms of Risks

Projects differ substantially. Figure 6.1 illustrates this by positioning various types of projects according to the intensity of the social/institutional, technical and market-related risks they pose to sponsors. For instance, *oil platforms* are technically difficult, but they typically face few institutional risks because they are often built far from public attention and are socially desired because of the high revenues they bring. Selling oil is not too difficult though prices can exhibit volatility. *Hydroelectric-power projects* tend to be moderately difficult insofar as engineering is concerned, but very difficult in terms of social acceptability. *Nuclear-power projects* pose high technical risks but still higher social and institutional risks. *Road and tunnel systems* present very high levels of risk especially if they are built by private investors. Rock formations usually hide big surprises. Difficulties in social acceptability abound when user fees are applied. Market risks faced by roads, bridges, and tunnels are very high when private sponsors under concessionary schemes build them. *Urban transport projects* that meet real needs pose average market, social, and institutional risks. However, they still pose technical risks, as they regularly involve underground geological work.

Research-and-development projects present scientific challenges but face fewer social acceptability and market difficulties as they can be broken into smaller testable investments.

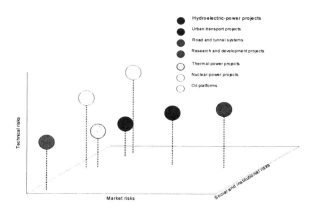

Figure 6.1   Projects differ substantially

6.3 The Nature of Risks in projects

Risk is the possibility that events, their resulting impacts and dynamic interactions may turn out differently than anticipated. Because of the nature of LEPs, practitioners tend to stress the negative impacts and take less notice of positive ones. Risk is typically viewed as something that can be described in statistical terms, while uncertainty applies to situations in which potential outcomes and causal forces are not fully understood. Both risks and uncertainties will be referred as risks. Risks need to be unbundled for clear understanding of causes, outcomes, and drivers.

In the IMEC study, managers were asked to identify and rank the risks they faced in the early front-end period of each project (Miller and Lessard, 2001). Market-related risks dominated (41.7 per cent), followed by technical risks (37.8 per cent), and institutional/sovereign risks (20.5 per cent). Figure 6.2 gives the frequency of citations of the risks that managers identified as important in their projects. Below, we will discuss in detail each class of risks as well as two additional topics: emerging turbulence and the oversight of opportunities.

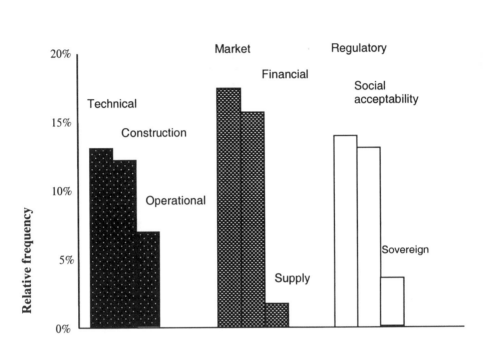

Figure 6.2   Major risks in large engineering projects

Market-Related Risks

The ability to forecast *demand* varies widely, thus creating high levels of risks. The output of oil projects is a fungible commodity sold in highly integrated world markets: probabilistic forecasts are possible. In contrast, road projects serve a specific area and customers who often have many options. Hence, it is difficult to forecast how the potential users of a highway, tunnel, or bridge will change their patterns of travel after the project is in place.

The market for *financial inputs* depends on prior risk management. Unless all risks have been addressed by sponsors, financial markets are hard to reach. If a project offers an adequate prospective return, it is often unable to go forward because of the parties' inability to work out acceptable risk-sharing arrangements. In addition financial markets are subject to cyclical processes such as the Third World debt crisis in the early 1980s, the 'real estate' crisis of the late 1980s and the Asian crisis of the late 1990s *Supply risks* are similar to market risks: both involve price and access uncertainties. Supply can be secured through contracts, open purchases or ownership but this sometimes results in higher costs than the current market prices during the exploitation of the project.

Completion Risks

Projects face *technical risks* that reflect their engineering difficulties and novelty: some of these risks are inherent in the designs or technologies employed. *Construction risks* refer to the difficulties that sponsors, prime contractors, and contractors may face in the actual building of the project, including the possibility of a lack of adequate competencies on the part of the contractors, or the potential for inter-organizational conflicts. *Operational risks* refer to the possibility the operation of the project will be deficient. Hence, the project will forego potential revenues and making it work will cost more than anticipated: such risks can be reduced substantially by the selection of an operator with an economic interest in enhancing revenues and controlling costs.

Institutional Risks

The ability of projects to repay debts and investments depends on law and regulations that govern the appropriability of returns, property rights, and contracts. Some countries are governed under constitutional frameworks and the rule of law, while others are led by powerful political parties or clans. Institutional risks are typically seen as greatest in emerging economies – in countries whose laws and regulation are incomplete and in a state of flux.

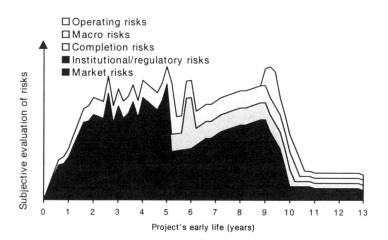

Figure 6.3   Dynamic evolution of risks

Regulations concerning pricing, entry, unbundling, and other elements are presently undergoing major changes in many countries, thus opening opportunities. *Social-acceptability risks* refer to the likelihood that sponsors will meet opposition from local groups, economic-development agencies, and influential pressure groups. *Sovereign risks* in turn involve the likelihood that a government will decide to renegotiate contracts, concessions, or property rights.

Turbulence

Many risks can be anticipated and addressed in the front end period. Some risks are linked to the life cycle of the project: regulatory risks, for instance, diminish very soon after permits are obtained and technical risks drop as engineering experiments are performed. However, other risks emerge over time and may catch sponsors and participants by surprise. Projects that appeared sound at a point in time all of a sudden become ungovernable. Turbulent events burst out and interact. Figure 6.3 illustrates risks that emerge and challenge sponsors.

Table 6.1 Examples of turbulent events affecting projects

Type	Example
Exogenous events	
Socio-political and macroeconomic	Financial crises (country or world) Major legislation (unexpected)
Unexpected natural events and discoveries	Catastrophically bad weather Unforeseen geology Discovery of valuable natural resources
Direct opposition to project	Court challenges by pressure groups Organized community opposition International opposition
Sovereign behaviour	Rule changes by regulator Refusal to grant permits Expropriation battles Granting of competing concessions
Market	Unexpected shifts in the patterns of use Abrupt changes in input prices (oil, gas, etc.)
Endogenous events	
Coalition unravelling	Withdrawal or bankruptcy of major partners Opportunistic moves Difficulties experienced by one partner
Uncontrollable interactions	Unexpected consequences of strategies Social deadlocks Accidents, strikes Complementary work not ready Contractor bankruptcy Problems with new technology, site, etc.

Hence, projects and their futures cannot be specified in advance. It is impossible to bring all parties to the decision table, gather all pertinent risk estimates and conduct all-encompassing negotiations to arrive at a final agreement that will fix the project and its context for the foreseeable future. Not only do new parties enter decision processes but the context changes unpredictably.

Turbulence is thus a series of 'surprises' to decision makers that introduce discontinuities and deeply felt differences between the future as it was imagined and the present reality. Both exogenous and endogenous events cause surprises. Table 6.1 gives examples of unexpected turbulent events (in the IMEC research) that temporarily stopped projects, required reassessments of viability, or demanded restructuring.

The most striking phenomenon observed by IMEC was the tendency of projects facing turbulence to enter spirals of disintegration. In the face of difficulties, parties have a tendency to leave projects or minimize their losses, perhaps at the expense of other participants. The inability to counter centrifugal forces leads to failure. Moves and countermoves lead to a vortex that causes project demise. Without a set of institutional and governance devices, viable projects enter into deadlock or demise.

Oversights

Oversights refer to the risks of loosing a good opportunity to improve value or to reduce costs. Oversights refer to the rejection by mistakes or inadvertence of an opportunity that could have generated substantial gains. Oversights do not hurt when the sponsors remain unaware of what could have been achieved. When however, there is a consensus that too many opportunities are lost, searches to develop better decision processes so that sponsors will not remain ignorant about what could be done.

Oversights risks are particularly important when projects are built using the traditional mode of contracting in which sponsors define in detail their expectations and call for bids for execution. By contrast, new modes of bidder selection which rely on partnerships or relational contracts are said to trigger innovative solutions that are much superior to those emerging from the traditional mode of contracting.

6.4 Approaches to Managing Risks in Large Engineering Projects

To help managers structure and cope with risks, theoretical perspectives can range from narrow reductionist analysis to systemic institutional approaches. In the course of the IMEC Program, we have observed that sponsors strategize to influence outcomes by using six main risk-management techniques: Decisioneering to assess and mitigate risks; Building robust strategic systems; Diversify risks through portfolios, Instilling Governability; Shaping Institutions; and Embracing risks.

Figure 6.4 illustrates these techniques along two axes: the extent to which risks are controllable and the degree to which risks are specific to a project or inherent to the economic systems thus affecting large numbers of actors. When risks are specific to the project and controllable i.e. endogenous, the prescription is usually to mitigate with project risk management approaches. In contrast, when risks are specific but outside the control of any of the potential parties in the project, shifting or allocating those risks using contracts or financial markets is the appropriate solution. When risks are poorly defined and under the control of affected parties, governments, or regulators, transforming them through institutional influence is the way for sponsors to gain control. When risks are broad, systemic, but controllable, the approach is to diversify exposure through portfolios of projects, hedging or insurance. Residual, systemic, and uncontrollable risks have to be embraced by sponsors.

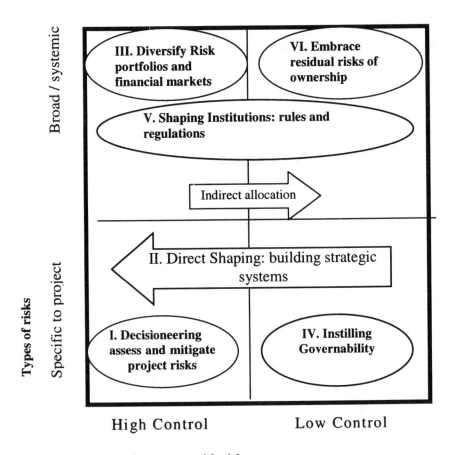

Figure 6.4 Strategies to cope with risks

6.4.1 Decisioneering Approaches to Assess and Control Risks

Decisioneering approaches view projects *as initiatives that can be planned in advance and under conditions of calculable risks.* Careful analyses of trade-offs between costs and risks, it is claimed, can yield good approximations for the appropriate timing of investment in projects. Accelerating a project will increase development costs to the point that there is a danger of sinking it. Proceeding with prudence increases the danger of missing the opportunity that the project aims.

The project-management field has been oriented for many years in this direction. Much of the accumulated knowledge appearing in the PMI Body of Knowledge falls in this category. This perspective assumes an environment in which the range of issues facing a project is more or less constant and current quantitative trends can be easily extrapolated into the future.

Decisioneering usually involve quantitative sensitivity analyses to investigate the impact that possible deviations in some variables, such as anticipated costs, can have on financial performance. Probabilistic approaches, using scenario analysis, decision trees or influence diagrams, provide more sophisticated alternatives to sensitivity analysis. Sensitivity analysis and similar approaches are helpful for making go–no-go decisions by eliminating the projects with high anticipated performance variability. However, because they focus on aggregate variables, they are less useful for the concrete and detailed shaping of a strategic system.

6.4.2 Building Robust Strategic Systems

Sponsors of projects deal with anticipated risks, constraints, and issues by creating strategic systems that integrate various elements with the goal of reducing the overall risk. A large portion of the risks are addressed with project-specific strategies, to reduce the odds of negative events or the maximal negative impact that such events can have on the project (Floricel and Miller 2001). We identified five classes of strategies summarized in Table 6.2.

i) Information/selection strategies refer to the approaches that managers use to gather information about the project and its environment as well as to shape and approve the project concept, identify and decide on the best strategies. We identified three classes of information/selection strategies: 'studies,' 'private search' and 'relational probing.' 'Studies' refer to 'impersonal' and 'objective' information gathering approaches such as literature search, forecasts, tests, or simulations. In this class, selection emphasizes decision theoretical models and bureaucratic procedures. The 'private search' class involves the use of a network of personal contacts to obtain 'privileged' information; it often requires a history of previous joint work and trust. Selection takes the form of early commitment to and relentless but flexible pursuit of a sole opportunity. 'Relational probing' refers to lengthy face-to-face interactions with potential participants, such as banks, regulators, clients, suppliers, engineering and construction firms, operators, and affected parties, during which the information emerges and concept is directly tested. Like 'personal search', 'relational probing' strategies use 'rich media' but they privilege meeting potential opponents and critics of the project, rather than friends, in order to proactively uncover flaws or risks. Selection relies on iterative discussion and negotiation to expose unworkable alternatives and stimulate the emergence of a better project concept. Practices such as inviting representatives of the public into planning and design meetings and proactively consulting conservationist groups and environmental regulators help to find credible solutions and reduce the likelihood of protest. The information/selection approach that will be used will influence the extent to which risks are identified and the quality of the solutions and strategies that will be produced.

ii) Co-optation strategies secure a basic set of 'core competencies,' such as technical and construction skills, which will increase the odds for success in critical areas of project execution and ensure access to 'resources' such as markets, financing, and even public support. The first step in co-optation is deciding which resources can be provided by business units or subsidiaries and which areas require bringing independent participants on board. With respect to external organizations the developers must decide the kind of link they will have to the project. Some participants can be brought in through 'partnership' links – as co-owners, joint-venture partners, or equity investors. Alternatively, resources can be co-opted through contracts and formal agreements such as project financing and tax treaties.

Traditionally, an engineering firm designs a project under a cost-reimbursement contract, and construction is contracted to a large number of firms using fixed-price or unit-price contracts. More recently, engineering-procurement-construction and turnkey contracts group these activities together to better align incentives between engineering and construction. BOT-like schemes, which make a single firm or consortium responsible not only for developing, designing, and building the project, but also for operating it for a long period of time, propose an even more radical way of aligning incentives. The kind of participant selection procedures used, for instance invited negotiation as opposed to an open and public call for bids, the detail of contract specifications are also part of cooptation strategies. Then again, access to some resources can be achieved through informal 'engagement' links with communities and other stakeholders in order to obtain their support. Another decision concerns the extent to which different resources and tasks determine the number and quality of participants, the structure of links between them and the bundle of resources, activities and decisions each one will control.

Understanding Risks and Shaping Large Engineering Projects   111

Table 6.2   Devices used in building strategic systems

Information search	Research and studies
	Personal search
	Relational Probing (e.g. early involvement of financiers, operators, and potential opponents)
Network building and co-optation	Partnership (Alliance of owners sharing equity and partnerships with suppliers or contractors)
	Contracts (e.g. Public–private partnerships) and methods of contractor selection (e.g. bidding vs. invitation)
	Engagement (e.g. Coalitions with project supporters or affected parties)
Structures of incentives, and contracts	Risks/decision rights allocation
	Type of price determination
	Incentives/penalties
	Frame agreements
	Contract flexibility, ability to restructure/contractual options
Project/design configuration	Geographical location/site selection
	Complementary investments and linkages
	Flexible/modular technical solutions
	Strategic structuring of activity and payment schedules
Influence and bold actions	Educate regulator, rating agencies, and others
	Side payments: compensation, add-ons
	Pre-emptive action, signal commitment
	Signal probity (e.g., bidding)
	Seek and improve on legal requirements
	Change laws and regulations

iii) Allocation strategies refer to the detailed ways in which rights, responsibilities, rewards and risks are apportioned between participants, through price, transfer, penalty, incentive and other contractual clauses. Parties to a contract delimit their respective responsibility areas; what each of them has to provide to the other party, when and under what conditions. A coal-gasification repowering project, realized by a joint venture between an utility, which owned the power plant to be repowered, and an independent firm that would build and own the gasification plant, provides an interesting example. The agreement between them stipulates that the utility provides a site for the gasification plant, guarantees the supply and quality of coal, demineralized water and auxiliary power. The owner of the gasification plant leases the site from the utility and guarantees the supply of synthetic gas and high-pressure steam. The utility has the obligation to accept all the synthetic gas that meets the quality requirements and owns all the by-products that result from the gasification process. Failure to supply reliably the required quantities triggers the payment of penalties.

The price-determination formula is another common contractual allocation strategy. While in cost-reimbursement contracts the owner bears the risk of a cost overrun, fixed-price contracts transfer the cost overrun risk to the contractor. In cost-incentive and performance-based price determination schemes the owners and contractors share the risks and rewards. In many power plant projects the price of the turnkey contract increased if the contractors delivered the plant early or if performance tests revealed that real plant capacity was larger than specified capacity. Other risk-allocation strategies limit the negative consequences for one of the parties to a contract. For instance, utilities often include 'regulatory out' clauses that allow them to cancel contracts with independent developers in case regulators do not allow them to fully recover the contract costs from their customers. Economic-dispatch formulas pass the additional costs resulting from operating a power plant at a sub-optimal capacity on to the electric utility that purchases the power and dispatches the plant.

Allocation strategies determine which participant will bear a specific risk but also the degree of influence the owners will have over contractors' actions and the kind of stance, adversarial or friendly, the contractors and owners will have one towards the other.

iv) Design strategies involve the use of technical, organizational, scheduling and financial choices to reduce the likelihood and impact of risks. One spectacular example of a technical solution used mainly for political risks is building a power plant on a barge that can be towed away from the coast of the host country in case of difficulties. Other examples are technical solutions that reduce the supply risk by providing fuel flexibility, repowering old plants to avoid regulatory risk, and design choices to gain the support of local communities. For instance, proactive discussion with regulators and the community led the managers of the Indiantown project to include in their project the construction of a 19-mile pipeline to bring cooling water from a heavily polluted canal rather than using local water sources.

v) Action strategies include confronting opponents using legal or informational means; persuading other participants and stakeholders such as banks, rating agencies, regulators, politicians, publics, and opponents; making gestures that legitimate the project in the eyes of the regulators or the communities; developing alternatives to be used if the preferred course of action is blocked by an adverse event; and taking pre-emptive steps to signal commitments. For instance, faced with the prospect of social opposition, the owners of the ITA power plant project, recently built in Brazil, established a public relations centre in the community and organized town-hall meetings in which the project was explained. Opposition weakened and the population became an ally of the project.

In the IMEC study, we have observed that the range of strategies used by sponsors in the shaping of their projects is statistically and significantly linked to the performance of projects. In fact the chi-square correlation coefficient between the scope of the strategic system and project performance is 8.3 with a confidence level 0f 0.015.

6.4.3 Portfolios, Insurance and Hedging

Some risks are systematic and linked to the world economic systems and countries' macroeconomic conditions. The usual way to protect projects against systematic risks is to use a portfolio approach. There are three perspectives within the portfolio approach.

i) Sponsors of risky projects likely to face turbulence build a diversified portfolio of projects to balance risks and cash flows. With a large number of different projects positive variations in a few prospects compensate for negative outcomes in others;
ii) Sponsors may hedge against possible losses due to currency fluctuations or commodity exposures by taking positions in futures for a fixed period of time or by investing in securities that has similarities to prospect;
iii) Sponsors may protect themselves against political risks by investing in many countries, finding partners in each country or buying insurance against expropriation.

6.4.4 Instilling Governability

In the last 20 years, the environments in which large-scale engineering projects, such as power plants, highways, bridges, tunnels, and airports, are developed have become increasingly characterized by turbulence resulting from shifts in institutional frameworks, political and economic discontinuities, a rise in environmental and social activism and, to a lesser extent, technological changes and innovations. Such changes clearly limit the validity of traditional risk management approaches.

Governing projects in turbulence is enabled by instilling four properties to projects. We have termed these properties respectively cohesion, reserves, flexibility, and generativity (Floricel and Miller 2001). As the four properties are often contradictory, a balance must be sought. For instance, strong inter-organizational bonds increase cohesion but limit flexibility. Conflicting objectives constitute one of the major obstacles precluding large-scale engineering projects from achieving governability in the face of turbulence. Hierarchical links create inefficiencies, while long-term contracts bring rigidities. Short-term contracts do not provide sufficient stabilization of the future to induce adequate investment. Increasing flexibility through design and incentives may sometimes reduce the efficiency of the project.

i) Cohesion is the property that results in participants staying with the project and solving the problems caused by turbulence, instead of exiting as crises erupt. The main sources of cohesion are the bonds between project participants resulting from co-optation strategies and informal links created during project execution or early operation. Still other bonds are the result of collateral ties between the organizations participating in a project.

A lack of cohesion leads to disintegration. This problem is well illustrated by the South Trunk project (real name disguised), an independent power project designed to burn waste coal from a nearby pile. During start-up, the project experienced repeated failures of the waste-coal and ash-handling systems. Participants started blaming each other. The owner blamed the turnkey contractor; in turn, the contractor blamed the owner because the coal received from the nearby pile had higher humidity than specified in the turnkey contract. An unexpected decline in coal prices ultimately led to the demise of the project. The payments that the project received from its utility client were tied to the cost of coal-fired power generation in the utility's plants. Technical difficulties became pretexts that each party jumped upon for an opportunity to exit with minimal losses. After making costly changes to the project and convincing the waste-coal-pile owner to change the contract and bring in higher-quality coal from outside, the bank took over the project but failed to operate it efficiently. The banks finally sold the project to the client utility for a quarter of its cost. The utility shut it down, arguing that it no longer needed the capacity, but later restarted it to face unexpected demand increases. Cohesion emerged quite unexpectedly as the basic governability property; one cannot govern a project that is disintegrating; flexibility is clearly not enough.

ii) Reserves of financial and other resources enable the project to respond during crises. Reserves can be built into the institutional arrangements surrounding a project. In fact, ownership is the dominant factor in building reserves. Traditional utility regulation in North America is built slack: utilities can use corporate reserves to salvage projects, and reasonable additional costs can easily be passed on to consumers. Co-optation and sharing, used to deal with anticipated risks, also build in potential reserves. In addition, reserves are frequently incorporated into execution budgets and schedules; contingency allowances in budgeted costs are a common practice for dealing with cost and schedule variability. Finally, reserves can be designed into projects through technical and other redundancies.

iii) Flexibility is the property that enables a project to restructure itself as choices, actions, and commitments, which initially stabilized the future, change when unexpected events occur. Flexibility can be achieved by using strategies that do not produce long-term constraints, offer alternative avenues for action, or reduce the costs of restructuring and pursuing alternatives. These costs can be reduced through co-optation and design strategies that emphasise modularity, in which no element of the project is critical. Contractual structures associated with co-optation and allocation strategies are one of the main sources of the lack of flexibility. The same long-term contracts that reduce market and fuel supply risks in independent power projects may block efforts to respond to new market realities. For example, the utilities that concluded power-purchase agreements in the mid-1980s under US PURPA Act found themselves locked in to paying high prices as prices dropped. They tried various means to change these contracts, but the only possibility was to exit by buying out contracts or buying and closing projects. Project-finance arrangements, frequently used by independent power projects, also create rigidities because of the numerous conditions attached. When specifications are not too detailed, turnkey contracts give contractors flexibility in selecting technical solutions, innovating, organising work, and so forth. However, such contracts often create rigidity at the interface between owner and contractor; as contractors stick to specifications, changes required by the owner will be very expensive.

iv) Generativity is the ability to develop creative responses to situations that appear difficult. Response generation presupposes correct sensing and interpretation, as well as the time and attention needed to generate constructive rather than destructive debates. Co-optation strategies, especially those that bring in participants with different competencies, may help. Having many points of view and access to different networks also means that adverse developments will probably be detected earlier and different perspectives will be brought into the discussion. For instance, unlike projects financed on the balance sheet, project financing brings banks, investment advisors, rating agencies, and consulting engineers to the heart of project debates. Creative individuals bring in new perspectives from outside the circle of managers who normally participate in the project. With their different experiences, they can sense danger and propose innovative solutions. On the other hand, numerous participants and contractual interfaces hamper creativity, especially when parties focus on contracts instead of problem solving.

6.4.5 Institutional Influence and Transformation Approaches

The prevailing framework of laws and regulations serves to reduce uncertainty and opportunism. In turn, new projects often present challenges and call for transformation of laws and regulations. The presence of coherent and well-developed institutional arrangements is, without a doubt, the most important determinant of project performance. Projects shaped in incomplete and shifting arrangements have a hard time taking off: they require deals and agreements that may not stand for long. In contrast, well-developed laws, regulations, and practices contribute significantly to enhancing project performance.

The main function of institutional arrangements is to help anchor projects in their economic and political contexts and ensure that investments will be repaid and social utility be provided. Unless they are solidly anchored, projects will be at the mercy of shifting interests, caprices, and opportunistic moves. Project anchoring requires three generic conditions:

i) Stabilization of the long-term future to enable investments. Legal and regulatory frameworks, such as private monopoly regulation and concession frameworks help to reduce risks by minimizing opportunities for clients, communities, or governments to attempt to capture revenues after the investment is sunk. The goal is to create the prospect of secure streams of funds in the long term to cope with the various uncertainties that can affect the project. To secure streams of revenues, the approach throughout most of the twentieth century has been to assign sponsorship and ownership to network operators. Recently, power-purchase agreements (PPAs), in which the regulator or the state forces a network to sign long-term supply contracts with independent producers, have been used as a tool for providing revenue flows. Concessions by the state to sponsors also provide a framework for future revenues but are less secure. Project practices such as purchase agreements, turnkey contracts, and project financing also reduce uncertainty by allocating risks among different types of participants.

ii) Flexibility to face turbulence. During the front-end development of projects, when agreements are negotiated and commitments made, managers develop specific strategies to cope with foreseeable risks; they cannot, however, develop specific ways to cope with 'surprise' events. Turbulence is likely to arise given the long time span required for development. Flexibility is provided by elements of institutional arrangements that enable projects to undergo rescheduling, restructuring, or bankruptcy. The flexibility provided by institutional arrangements helps many projects survive unforeseen events.

iii) Enhancing the legitimacy of projects, participating organizations, and agreements. Many projects face opposition from interest groups. Laws, regulations, and practices that create well-structured assessment frameworks enable sponsors and interest groups to air their views through public hearings, and even to oppose decisions through appeal procedures. Public-bidding frameworks structure the orderly selection of fitted doing frameworks structure the orderly selection of fitted sponsors and provide legitimacy.

Frameworks that structure public consultations and decision making, and regulate the trade-offs between conflicting interests make it possible to site public transportation systems, erect power plants, and, in some countries, build nuclear facilities. To manage social-acceptability risks in sitting of power plants in Japan, for instance, the Three Power Source Laws System was put in place by the Japanese Ministry of International Trade and Industry. This framework structures public consultations and hearings across the country; the population is consulted on the choice of eventual sites for projects and their technical features.

Institutional arrangements for shaping and building projects are prime determinants of success. However, sponsors attempting to anchor projects often find that laws and regulations are incomplete. Many projects serve to unlock new models of project delivery (for example, the first BOTs developed in the 1980s). Some projects are specifically used to experiment with and eventually implement policy changes (for example, the Tennessee Valley Authority experiment).

One third of the projects analyzed by IMEC required at least one change in laws and rules. Concession rights, property rights, economic regulations, or foreign-investment rules needed to be modified. More than one-fourth required or accompanied changes in property rights: land rights, water rights, monopoly of or improvements to BOT and concession frameworks. Changes to laws and regulations in capital markets were also frequent. A few projects called for new environmental frameworks.

Tremendous effort, in terms of both time and money, is required from sponsors to understand institutional environments within the geopolitical boundaries where their project is undertaken, so that they can align their strategies to this context, assess risks related to institutional change, and develop strategies to act upon them. Understanding prevailing institutions enables sponsors to assess manoeuvring space they have for undertaking a particular project.

6.4.6 Embracing Residual Risks in Shaping Projects

Diligent sponsors do not sit idle, waiting for the probabilities to yield a 'win' or a 'loss,' but work hard to influence outcomes and turn the selected initial option into a success. They shepherd their choice in light of changing conditions and often succeed against odds.

Projects are often described as games in which chance, force, or improvization dominates. Here are a few indicators of this untidiness. First, projects are often launched by promoters who need to convince and charm potential participants. High levels of expenditures need to be allocated to soft issues such as opinion research, public affairs, participation, and debates that appear to many executives to be money spent on non-business issues. Politics and power are more important than analysis. Decisions are never final but are remade, recast, and reshaped until the project is feasible. Confrontation not only overtakes collaboration but often brings deadlocks. Crises punctuate orderly development and create a climate of disorder for those who believe that decisions ought to be made in a deliberate fashion.

6.5 Shaping Projects Through Iterative Planning

Planning may be portrayed as ineffectual (Mintzberg, 1994), but, paradoxically, sponsors are spending increasing amounts of resources and time on this activity. No sponsor would agree to reduce investment in planning, analysis, and simulation because the future is unpredictable or unknown. Resolution of the ambiguities of needs and solutions is usually achieved through successive redefinition, bringing the project toward committable closure. Many project concepts will have been killed early, mid-course, or at the end.

Competent sponsors focus on shaping difficult projects that have potentially high payoffs. Using a mountain-climbing metaphor, competent sponsors do not rush to climb the tallest mountains or only the mountains that they are best equipped to climb. Rather, they seek to select, equip, and train a climbing party that should be is developed, even though these cannot be fully specified in advance. In fact, the game consists of identifying projects that stretch the limits of the firm's capability but that, because of their complexity and risk, offer substantial benefits to clients that cannot be achieved with simpler, less risky undertakings; these benefits have high appropriability, since few other firms will have the capabilities to exploit them. If we trace the development of leaders and the projects that they sponsor over time, we see that they have regularly 'pushed the envelope' to identify projects on the frontier of their capability but not yet commoditized.

Projects are shaped in episodes to transform the initial hypothesis, make progress on issues, and solidify initial coalitions of players, in order to achieve temporary and, eventually, final commitment. Each episode opens new options and closes old ones until sponsors and partners achieve final lock-in, thus binding their commitments and losing most of their degrees of freedom. Shaping episodes start with momentum building, continue with countering opposing forces, and end with closure. Figure 6.5 pictures shaping efforts as going up a hill through coalition building, problem solving, and risk management in the face of counter-dynamics such as cynicism, false expectations, and feedback effects.

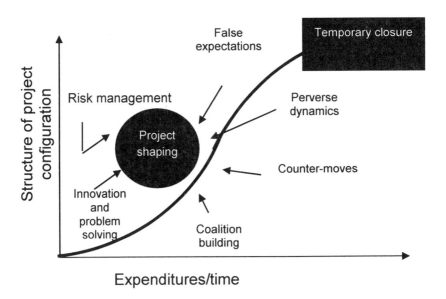

Figure 6.5 Shaping efforts

Momentum is built by imagining solutions to face risks directly or indirectly, promote legitimacy, and design a project configuration such that partners, affected parties, and governments believe what is proposed. Outside parties commit to credible projects sold to them optimistically by reputed sponsors. Coalitions are built through mutual accommodation in which the parties exchange and agree to reciprocal non-zero-sum games. Shaping efforts aim to resolve indeterminate situations through bold commitments and leveraging of pertinent factors to arrive at agreements that are optimistic but not delusional. Eventually, imperfectly coordinated but stabilized understandings move toward focal points at which temporary agreements become enforceable.

Countering forces such as doubts, cynical actions, and fear limit the effectiveness of shaping efforts and may even plant seeds of later failure. Sponsors sometimes believe their own overly optimistic assumptions. Weak analyses, incomplete market research, and the search for contracts lead to the selection of erroneous paths. Early lock-in thus closes off reconsideration of project configurations and downplays future options. Excessive realism, in contrast, leads to scepticism and the eventual rejection of good opportunities. What is basically a good project is painted negatively by some and rejected by others. Leaders lacking in shaping experience and credibility cannot communicate believable expectations. Unfavourable judgments regarding events, exposures, and design choices drive away parties whose contribution is critical. Doubts, negative stories, and emergent problems set in motion self-fulfilling prophecies.

Sponsors often yield to the temptations of unreasonable commitments. Clients, bidding frameworks, or business relationships demand that project solutions be developed fast. Sponsors accept risks, hoping that the downside will never materialize, but winner's curse often sets in. Leaders and sponsors can become blind to particular risks and develop no mitigation strategies. Blindness generally comes from the inability to form coalitions that include partners with the relevant viewpoint. Parties that have committed to behave in specified ways fail to assume their responsibilities.

Games between agencies and sponsors will take many forms: refusal to grant permits, delimitation of project conditions by regulators, changes of rules during project construction, legislative changes against which no action is possible, and absence of support when crises arise. Perverse dynamics can take hold when inefficiencies and feedback are not countered by further strategic efforts. Only projects whose leaders and sponsors who have the resources, willingness, and competencies to respond to destructive forces survive.

Closure: Each shaping episode ends with a process of closure that opens new options, suggests abandoning the whole project, or calls for temporary agreement on a project configuration or final lock-in. Closure is a shared agreement between the leaders, sponsors, and key players that the original hypothesis about progress on solving issues is confirmed. When final closure is made, proceeding to engineering, procurement, and construction is the next step.

Closure takes many forms as sponsors progress through multiple episodes: memorandum of understanding, business case, negotiated agreement, formal public commitment, sets of formal contracts, and so on. The dangers associated with closure are that choices can be made too early, too late, too rigidly, or too flexibly. Missing the boat—rejecting a good opportunity—is just as real a possibility as selecting a bad option or pursuing the wrong project. Premature closure locks a project on a rigid configuration, narrow sets of agreements, or irreversible choices that limit degrees of freedom for the future. Generative closure, in contrast, is the selection of a temporary project configuration that opens a new hypothesis, triggers new options, and retains degrees of freedom for later actions.

6.6 Conclusion: Creation and Exercise of Options

The front-end process to cope with risks can be re-interpreted in terms of the real-options framework that is currently revolutionizing academic treatments of project evaluation. In fact, as is often the case with cutting-edge practice, managers have been successful at creating value through the development and exercise of sequential options without explicitly framing the process in options terms. Academics have simply codified this practice in the form of a new conceptual framework.

The real-options framework is based on the same logic as that of financial options as developed by Black and Scholes (1974). Dixit and Pindyck (1995) and Trigeorgis (1996) extend it to real options, while Laughton and Jacoby (1993) provide operational specification for options valuation. Kulatilaka and Lessard (1998) have demonstrated how the real-options approach can be combined with a decision-tree framework.

The real-options approach recognizes that decisions that determine project cash flows are made sequentially over many episodes. The key insight of this approach is that uncertainty or volatility can actually increase the value of a project, as long as flexibility is preserved and resources are not irreversibly committed. As a result, the economic value of a project when it is still relatively unformed is often greater than the discounted present value of the expected future cash flows. Value is increased by creating options for subsequent sequential choices and exercising these options in a timely fashion. Thus, sponsors seek projects that have the potential for large payoffs under particular institutional and technical circumstances.

References

Black, F. and Scholes, M. (1974), 'From Theory to a New Financial Product', *Journal of Finance,* vol. 29, no. 8, pp. 399–412.

Dixit, A. K. and Pindyck, R. S. (1995), 'The Options Approach to Capital Investment', *Harvard Business Review,* vol. 73, no. 3, pp. 105–16.

Floricel, Serghei and Miller, Roger (2001), 'Strategizing for Anticipated Risks and Turbulence in Large-Scale Engineering Projects', *International Journal of Project Management,* vol. 19, no. 8, pp. 445-455.

Kulatilaka, N. and Lessard, D. (1998) *Total Risk Management,* Working paper, Sloan School, Massachusetts Institute of Technology, Cambridge, Mass.

Laughton, D. G. and Jacoby, H. D. (1993), 'Reversion, Timing Options, and Long-Term Decision-Making', *Financial Management,* vol. 22, no. 3, pp. 225–24.

Miller R. and Lessard D. with S. Floricel and the IMEC Research Group (2001), 'The Strategic Management of Large Engineering Projects: Shaping Risks, Institutions and Governance,' MIT Press, Cambridge, Mass.

Mintzberg, H. (1994), *The Rise and Fall of Strategic Planning,* Free Press, New York.

Shapira, Z. (1995), *Risk Taking: A Managerial Perspective*, Russel Sage Foundation, New York.

Trigeorgis, L. (1996), *Real Options: Managerial Flexibility and Strategy in Resource Allocation,* MIT Press, Cambridge.

Chapter 7

The Government Role – From Market Failures to Social Capital

André Manseau and Elisabeth Campagnac

Construction is a major industry in every nation. Its economic and social impact is very significant and pervasive, and this context often urges strong government intervention for the sector and/or for its customers. Construction and related sectors (including manufacturers of building products and systems, as well as services sector firms involved in construction activities such as architects, engineers, designers and property managers) account for about 14-15 per cent of the national GDP. It also has a significant impact on living standards and on the capability of a society to produce other goods and services and to trade effectively. Its products largely define the urban environment.

Governments also own and operate an important share of the built environment in every nation. Governments of the OECD group own between 10 and 25 per cent of their total country fixed assets in a form of public works, which include buildings and general purpose infrastructure (OECD, 1997).

It is generally agreed that construction is an industry that plays a key role in the creation of the asset base of a country. It can be reasonably assumed that the public policy regime for this sector will reflect the political, social, economic and cultural values of a given country.

7.1 Economic Paradigms and Government Role

After World War II, governments from many industrial countries invested heavily in construction for rebuilding destroyed cities, roads and infrastructures, as well as for supporting a booming population. This situation persisted for almost thirty years, up to the mid 1970s.

Interestingly, this significant public intervention occurred in an economic period of strong neo-liberalism. The neo-liberalism or 'laissez-faire' paradigm broke through around 1900 and was reinforced with the Schumpeterian model and the Chicago School. The entrepreneur is considered the driving force, creating firms and business activities, with the objective of maximising his or her profits by using different inputs (capital, labour, land and material) at the lowest possible cost for producing outputs (goods and services) at the highest possible price. The basic assumption for creating general or national wealth was to let these champions go wherever they want, by free trade, 'laissez-faire', and reducing barriers of any kind (regulations, tax, resources access, etc.), they, in turn, become wealthy and their wealth is distributed.

With the 'laissez-faire' paradigm, governments limit their roles to closing some economic gaps, called 'externalities' or 'market failures'. However, it appears that there have been many market failures in construction – therefore important roles for government. Amongst the externalities in construction are public safety, energy conservation and environmental protection. Construction products have impacts beyond the direct client, affecting the community as well as future generations. These externalities are often described as creating social benefits with the help of public investment. For example, regulations have been effective in the case of energy efficiency in housing and have significantly reduced energy consumption (Gann, 1998).

A new type of market failure has been stressed in recent years, with the strategic importance of knowledge. It is called 'asymmetrical knowledge'. Some firms or individuals have access to more knowledge or are faster than others in acquiring strategic knowledge. It provides them with competitive advantages, as they can capture a significant share of resources or market before any competitors have even realized it.

In such a situation, it could be considered good public policy to bring in measures for protecting 'less smart' firms or consumers. This is done in the construction industry, particularly to protect buyers; there are many mechanisms in various countries to this end. We can have inspectors, mandated or licensed by government, for inspecting buildings (for example in the Netherlands); a state licensing system of market providers or workers (Japan, Germany) or a law imposing a 10-year guarantee of construction (France). Building codes, although not limited to that role, are also seen as a mechanism for protecting consumers. Finally, a state can have regular awareness programs for educating and informing the consumer.

Another way of presenting externalities in construction has been described by Seaden (1996), based on Paul Romer's works. The author developed a matrix (Table 7.1), presenting construction goods in two axes: by its degree of ownership control (from private to public goods) and by its capability of being simultaneously used by different people; from 'rival goods' to 'non-rival goods' (not affected by their use).

Table 7.1   Type of construction goods (Adapted from Seaden, 1996)

	Rival	Non-Rival
Private ownership control	Land, private building	Patents, copyright
Shared ownership	Shared equipment or facilities within a group or consortium	Best practices developed through partnership
Public	Waste water systems	Published or accessible documents and information (publications, books, Internet)

The 'neo-liberalism' approach suggests that government's role is essentially limited to two types of interventions: to own public goods and to establish rules for protecting the private owner of non-rival goods. However, it is not so simple, as governments also act by defining private ownership of many goods (and perhaps implicitly for all types of goods), and thus they define attributes private ownership and rights on goods. Land, a building or a patent must be registered in order to become private goods. Any dispute between private owners must be refereed.

Therefore, we may say that government can intervene in a number of different situations and that its role is not based on economic paradigms or theory alone. The government's role appears to be more as an evolving social relationship between the state and its people. Any society or community builds rules, co-operative activities and standards. Standards and rules are therefore 'social constructs' (Amable et al., 1997). There is no more 'pure' role but roles for addressing 'minimal societal values'. These roles for addressing minimal societal values, as well as some other important values, are many kinds of interventions make possible (Seaden, 2002).

Free-market is a concept that cannot stand by itself. Any market and trade needs rules for:

- Avoiding stealing and robbing;
- Protecting freedom and access;
- Respecting each other;
- Refereeing disputes.

Indeed, an increasing level of trade activity creates the need for more rules, procedures and inter-mediating support. '…markets do not regulate themselves. No policy is also a policy. Globalization does not invite an era of laissez-faire, but a thornier set of questions about governance' (Kuttner 1997:30).

A socio-economic approach would better explain the relationship between a state and its nation. The evolutionary and institutional economics could certainly shed some lights for describing and explaining state and institutions changing roles. With this framework, policy formulations and implementations are an ongoing development and improvement of negotiated standards or rules of behaviour. Any standards or collective agreements can be continually challenged for a perceived better one.

Evolutionary economics explains economic growth as largely the result of the progressive introduction of new technologies or innovations that provide improved processes or goods. However, there are no optimal, but path dependency and satisfactory processes, called 'routines'. Physical technologies continue to play a leading role, but physical and social innovations co-evolve. A routine is a way of doing something and could be a way of organizing work, a kind of market transaction, a law or a policy. A routine is usually linked with other routines, tends to be the result of the cumulative contributions of many parties, and takes place over many years (Nelson and Winter, 1982; Nelson and Nelson, 2002).

There are several implications at the policy level of such an evolutionary and institutional perspective. The motor of the economy is no longer the individual firm alone, but the performance of the overall economic system, constituted of firms and all other organizations is evenly important. Innovation and learning are at the centre of economic and social development. Information is incomplete and in part tacit and embodied, implying that performance is contextual and never optimal. There is no spontaneous or radical change. Changes occur among interdependent players who learn from each other and therefore evolve together through particular paths.

Developing efficient systems where individuals can fully participate and benefit is not always an easy task. Organizations, just like individuals, work in networks with multi-partners; they learn from each other and adapt themselves in order to develop their unique but complementary expertise. They perform by building coalitions and agreements in order to develop widely recognized and desired competencies or goods that become dominant or standard in the market. (Saviotti, 1995; Foray, 1993).

Rules and policies are socially relative and ever changing. Accordingly to many observers, government's role is evolving toward building social capital. Social capital has been defined by Portes (1998) as 'the ability of actors to secure benefits by virtue of membership in social networks'. Individuals use social ties to create reciprocal obligations to access strategic information, to decrease transaction costs, and to increase their bargaining power (Blyer and Coff, 2003). Social capital consists of the norms, roles, institutions and networks of social organization (Dasgupta, 2002).

These norms and networks provide individuals and communities with resources and support. However, similar to other forms of capital, valuing an asset required to take into consideration the use (or potential use) of the asset – to make it works and benefit from this investment (PRI-SSHRC, 2003).

Because of a variety of contextual factors, different socio-economic systems are possible. The continuous search for improved processes or policies requires diversity creation as well as a systemic approach. In a global world, national governments need policies for developing a desired uniqueness of their firms and organizations within the international market. Furthermore, rules and standards have to be quickly and effectively developed in a quickly changing world.

7.2 Co-Existence of Different Systems

The concept of National Innovation Systems has been used in a number of OECD studies in the 1980-90s (OECD, 1996; 1997). Based on the evolutionary and institutional perspective, an innovation system is defined as the network of organizations, firms and institutions that interact and learn in developing innovations. Taking into consideration that a large part of these interactions occurs within national borders (which might be less true with globalization), each nation develops its own institutions, rules and laws, and so, creates its own national system of innovation.

Beyond national identity, some countries appear to have similar features. Amable et al (1997) identified four major types of 'social systems of innovation' amongst the highly industrialized OECD countries:

- The Market-driven system, present in the USA, UK, Canada and Australia, assumes that the market allocates resources in the most efficient manner through the bidding process and that government, even though a significant buyer of construction goods and services is just another participant in the market place. Construction practitioners in these countries display significant distrust in government's ability to influence economic development and they believe that public intervention should only take place in the case of an obvious market failure. Regulation is to be kept at the minimum, dealing primarily with safety and consumer protection. Labour arrangements are flexible and workers are highly mobile. There is an assumption that innovation occurs due to opportunities created by the competitive forces in the market place and by systematic deregulation. Public sector actors are considered inherently conservative in their acquisition practices, resisting innovation. Commercial negotiations are adversarial in nature, creating concern over liability when new practices are introduced. Government sponsored demonstration projects are sometimes used to stimulate innovative ideas, which are then disseminated to the industry;
- The Government-led system, which can be observed in France, Germany, Italy or the Netherlands, sees government, with its large purchasing force and social responsibility, as playing a central role in the market place. Publicly sponsored projects are often used to initiate or demonstrate new

technologies that are then disseminated to other practitioners. There is a significant level of regulation in all aspects of construction. Labour arrangements tend to be inflexible with little mobility. Public policy instruments are seen as essential elements of innovation, with government being perceived as a valuable partner. Commercial negotiations tend to emphasise existing linkages and government may intervene to achieve the desirable socio-economic goals;
- The Social-democratic system, more prevalent in the Scandinavian countries, shares many of the characteristics of the government-led system with a particular emphasis on the tripartite (industry/labour/government) approach of solving industrial issues. Labour arrangements show a greater level of flexibility and consideration of competitive forces. Innovation is state supported, and is expected to balance social, economic and environmental values;
- Finally, the Meso-corporatist system, typical of Japan, is based on the presence of very large corporations. Domestic competition in construction is intensive but a certain amount of market sharing exists, which has allowed individual companies to achieve relatively high margins. Government expects reinvestment of excess profits in innovation, because innovativeness is considered as important corporate and national value. Public policy is focussed on supporting large companies expected to lead in continuous technological and quality improvements. Labour flexibility within organizations is relatively high in a context of lifetime employment. Public policies on technology directions, regulation, training and innovation related issues are arrived at through consensus negotiations between public sector and major industrial participants.

For construction, there are significant differences between national attitudes towards the role of the government as the regulator or the principal customer in the liberalization of domestic markets, in labour relations, education and training, in the legal regime and in the methods of financing (Manseau and Seaden, 2001):

- Countries with a more centralized government structure (Japan, France, United Kingdom, Netherlands, Denmark, and Finland) always have a national 'construction' ministry that has been able champion the particular needs of its constituency and promote customized innovation-enhancement policies in construction. There is evidence of public concern with specific characteristics of the construction industry and of recent policy changes. Research funding is being redirected from products to processes, government acquisition practices are being modified to stress value over price, industrial collaboration is being promoted and technology/knowledge brokers are being introduced;
- Countries with a federal type of constitution (USA, Germany, Canada, and Australia) and a more decentralized government structure tend to

place the responsibility for construction at the state (province, Lander) level. Generally, there is no central focus or champion for the construction industry; instead there is a multitude of various agencies dealing with particular concerns such as acquisition of public works, technology development, safety, consumer protection or losses due to natural disasters. For example, we may find a number of government supported agencies for the evaluation of new products and systems. As well and possibly as a consequence of the lack of a national point of convergence, there is no unified industrial representation of the construction point of view to the senior levels of government. There does not appear to be significant public policy interest in construction innovation separately form the general issue of industrial development. Government can also provide innovation leadership, incorporating new products/systems in its building programs and/or providing unbiased public assurance as to their 'fit for use' and thus enabling more rapid introduction and commercialization;
- The construction industries in less industrially developed countries with significant economic challenges face important problems of providing large volumes of new housing and infrastructure at a reasonable cost. The maintenance-renovation component is lower, in comparison with more industrially developed countries. Governments in these countries are usually the most important clients for construction, but public policies towards innovation do not appear to address problems related to infrastructure development. Few programs to support innovation and/or R&D are offered and they are often based on public policies for the high-technology sectors in wealthy OECD countries.

No one single national system has appeared superior in all aspects. We identified a number of different interesting features in each country (Manseau & Seaden, 2001). Therefore, different systems may not only co-exist, but each provides each some interesting socio-economic impacts. For example:

- Japan's construction expenditure and its investment in R&D are the highest of OECD countries. It has developed, through historical evolution, several large, vertically integrated construction companies that believe in technology as a major competitive advantage. Large, in-house research institutes support corporate activities. Thus, in contrast with other countries, most of construction research is concentrated in the private sector. Government has been concerned with the future of Japanese construction and is currently promoting increased collaboration in R&D efforts; however, agreement on shared goals among the key stakeholders remains a challenge;
- Australia's construction expenditure per capita is average by OECD standards and its investment in related research, primarily in the public sector, is considered very low. Government is seeking to create a more

internationally competitive industry through its Action Agenda, which focuses on education, greater diffusion of technology and enhanced innovation. Collaboration among all actors is being promoted, with the public sector taking modest initiatives, while the private sector lags behind;

- The United Kingdom has had a sequence of high-level comprehensive reports during the last decade that examined its construction industry, and found it wanting. Several public/private innovation initiatives are now taking place (with the endorsement of senior politicians and corporate leaders) to achieve ambitious performance goals. Focus is on the supply chain, best practice and knowledge sharing. The public sector is encouraging innovation through proposed changes to its procurement practice, moving from 'lowest cost' to 'value-for-money' through business-like continuous improvement values;
- Denmark promotes innovation in a context of an 'organization society' which encourages socio-political negotiations on shared objectives. In the post-war period, it was able to successfully implement lasting public policy of general use of prefabricated elements. Currently, government-led demonstration projects attempt to develop process innovation that will achieve 'twice the value for half the price'. Results, so far, have been less than anticipated and more intensive coordination efforts of multiple stakeholders and of various policy instruments are required;
- Finland is considered as having a very effective innovation system (fifth best in the world) and it has recently experienced a high rate of economic growth. Public policy on construction innovation aims to strongly involve the end-users (real-estate sector) and to encourage rapid commercialization of ideas. There is a comprehensive array of public policy instruments to encourage innovation in construction at all stages of development; nevertheless, SMEs in the construction industry are perceived as lagging others, reluctant to obtain outside financing to expand their operations;
- The Netherlands' public policy stresses pre-competitive technological cooperation and special aid to SMEs. There has been a relatively positive response from the construction industry, with inter-firm alliances of 11 per cent (24 per cent for all sectors) and use of government innovation subsidies of 11 per cent (25 per cent for all). However, local enforcement of land use and of stringent environmental requirements is seen as inhibiting innovative practices.

However, despite the variety of public policies between countries, construction generally appears as a highly regulated industry. Although regulatory systems vary, regulations are widely used in managing building permits, construction codes, products and services certification, labour certification, procurement, fire risk, subsystems codes (plumbing, electricity, etc.), energy efficiency, the environment, etc.

In every industrialized country, the process of developing regulations and construction codes is complex, relying upon a number of evaluations, testing and consultations. A general convergence towards performance or objective based codes is enabling flexibility and adaptation to particular situation, but certainly not decreasing process complexity.

7.3 Common Trends – Globalization and Services

Contrary to the concept of co-existence of different national systems, a significant number of common trends were identified in a recent study of public policies in construction (Manseau and Seaden, 2001). The political/constitutional structure of countries and/or their social system of innovation did not appear to create radical differences in their national approaches to their respective construction industries. However, the government structure, the type of national innovation system and the existence of institution(s) clearly representing industry's needs does seem to influence the relative level of public concern with the choice of specific policy instruments.

The general situation, which can be observed in most of the fifteen countries studied and could be considered as the current general context for innovation in the construction industry, is as follows:

- National governments are showing increasing desire to encourage industry-led, longer-term goal setting, and to facilitate cooperation among construction firms for greater effectiveness;
- The most common innovation policy instrument is still the technology push model, through public funding of investigator-led research in universities or government institutions;
- Mission-oriented research policies are becoming less popular due to the perception that 'government does not always know best'. Directly funded construction research institutions are seeing their budget base eroded and have moved out of necessity and/or by policy directive into more collaborative arrangements with the industry;
- There is some evidence of construction related policies focusing on the management of supply chain and on-site problems. Emphasis is shifting away from product or manufacturing processes to organizational processes;
- The market pull approach, with industry initiated R&D and government providing support through cost sharing or tax relief is gaining popularity. However, response from the construction industry to this type of programs has not been encouraging, as a very low percentage of firms have taken advantage of these types of government aid;
- Information and communication technologies are perceived to have a significant impact on construction processes;

- Environmental issues and compatibility with community interests are increasingly influencing all construction phases, from initial site approval to eventual deconstruction and recycling.

The first five general trends above reflect those of innovation policies for all industries. A study prepared for Tekes, Finland (Frinking *et al* 2002), of innovation policies in seven countries (Germany, Ireland, Netherlands, Singapore, South Korea, Sweden and United Kingdom) identified central policy themes as: emphasizing public-private-university collaborations, stimulating development of SMEs, promotion of entrepreneurship, support for regional development, and the promotion of strategic research (more economic impacts expected).

Another major common change occurring in the construction industry is the shift toward a more service- oriented industry, which has significant impacts on public roles. With the increasing importance of managing existing stock, and addressing the entire construction life cycle and of developing more complex/sophisticated buildings, a number of new services have emerged or been significantly enhanced such as construction briefs, product designs, project management, facilities operations and maintenance, and project financing (Campagnac, 2001).

This industry reconfiguration can also be explained by the limits of the traditional (former) structure. The fragmented production system - divided in multiple phases (design, contract, sub-contract, monitoring, inspecting, etc.) - has created a number of problems and rigidities:

- Inability to innovate;
- Adversarial relation;
- Risks transferred to customers.

This trend toward a more service-oriented of industry involves an increasing role for the client, a flexible contractual arrangement and an emphasis on continuous communication, rather than on initial specifications. The change also has a significant impact on interaction between firms, increasing interdependencies and the need for a partnering approach. Finally, this industry reconfiguration is influencing a variety of public policies:

- Increasing role of communities as indirect or direct clients of large projects;
- Shift from descriptive to objective-based building codes;
- Use of a variety of public-private contractual arrangements for public construction, operations and maintenance; the private sector being considered as more of a partner in providing public services than of a product supplier.

A recent study on innovation in the service sector (CRIC, 2002) showed that government still has a significant role in new public-private arrangements for electricity, gas and water utilities. Public priorities and type of procurement greatly vary from one country to another. In some countries, large changes occurred in the structure of their 'public' works toward greater private ownership and concessions.

These common and rather international trends bring a new role for government, which is more of a supportive or partner type than dictated or declared. For example, in the UK, the government strongly supported and facilitated the development of national construction goals, as well as the establishment of a national construction industry association and implementation of a vast program, called M4I, for promoting innovations in construction (CRISP, 1997; Winch, 1999).

Trade liberalization under different international agreements has decreased dominance of domestic markets in construction. An industry consolidation is occurring, with an increasing number of large international construction firms. However, international activities are still limited to very large projects and accounts for a rather marginal share of the total market. As indicated in Chapter 2, the top 20 international construction firms accounted for US$ 123 billion in 1999, and the world construction market was estimated to about US$ 3.5 trillion that year (Flanagan in NRCC &CIB, 2001).

Trade by itself cannot explain common international trends in construction and in the role of government. There is still a low global integration in construction, from a trade point of view; however, as international standards, shared values and networks are being developed, major common trends are emerging in the industry. Culture and social values appear to flow much more easily than goods; this is where we can locate a large part of the explanation.

As mentioned in section 7.1, government role centres around its relationship with its people - developing 'social constructs' for addressing 'societal values'. Social constructs are developed from beliefs, but also with the help of trust that appears to be a key factor of developing social capital.

Shared social values, such as industry leadership in innovation, partnerships in innovation, environmental issues and community involvement are influencing the development of common public policies and approaches to governance in OECD countries.

7.4 Building National Social Capital in a Global Village

In light of the decreasing social distance between countries and of the development of common shared values, does national government still have a role? Is their role limited to harmonization, agreements and coordination with other states? If public policies reflect social values, are we moving towards a similar set of public instruments related to innovation and the built environment as social values are becoming more common internationally?

In construction, it has appeared in recent years that most of the currently available public policy instruments in support of innovation and improvement have not been of great use to the industry. Construction firms make a low use of R&D programs and their R&D investment is still very low. Construction is still facing major challenges with a high level of fragmentation, self-employed persons, lack of training and last minute management (Winch, 1999).

However, some approaches have had interesting results (Manseau and Seaden, 2001):

- Programs with greater local presence, focused on access to technology, and the promotion of collaborative arrangements, seem to be more successful. Furthermore, institutions that are able to evaluate new products or processes before market launch have also proven successful;
- Governments, as major buyers of construction services and as an interested party in building nation's sustainable capital, have succeeded in promoting long-term value and performance rather than the initial cost, with more open acquisition policies and public/private partnering arrangements;
- Regulations based on performance objectives, safety of occupants or users of buildings and infrastructure as well as compatibility (or a required balance) with community values and longer-term sustainability, have also had positive impacts on the industry;
- Governments that have stimulated and facilitated national discussion and reflection on industry issues and action plans have promoted a climate of trust and adoption of innovation and improved practices.

These approaches are also rather well aligned with the four suggested roles for government as identified in the Fairclough Report (2002): regulator, sponsor of innovation, client and policy maker. Considering that policies include actions in regulations, procurements or the facilitation of changes, we would suggest three types of interventions, as presented in Table 7.2, and how they can either act at the local, national or international level.

Table 7.2 Types of government intervention at local, national and international levels for supporting innovation and improvements in construction

Type of role	Local	National	International
System integrator	Mobilizing local strengths and efforts for developing national and international reach and strategic niche	Develop national strategies that make use of national resources as complementary assets and reinforcing each others	Facilitate access to international resources as well as transaction to trade and exchanges
Partner in sustainable solutions	Public-Private partnering for providing community services	Optimize national access of and benefits from public services and facilities	Select best suppliers and partners for national services and facilities
Participative standards	Facilitate active and comprehensive participation in elaborating standards	Establish national standards based on shared values	Harmonization and leadership in developing international standards

- System integrator/facilitator for developing collective action plans for positioning the industry as a leader (benefits to all citizens) and in a sustainable manner. A number of countries, such as the UK, USA, Australia and Canada have developed national construction goals and/or strategies by stimulating discussion and reflection as well as by facilitating consensus building on priorities (Winch, 2000; NSTC, 1999; Hampson & Manley in Manseau & Seaden, 2001). Systems facilitation also includes refereeing and facilitating multi-partnering arrangements, and systems integration involves addressing environmental issues and community values. Community strength is built through an ongoing connective and learning process, as well as by developing a competitive advantage in the global market. This role is rapidly changing with globalization, and it requires ongoing comparative analysis with other countries for developing and maintaining a strategic position. Further studies are required in this area, however;

- Partner with the private sector for providing community services, acting as a 'model' participant (evaluating and sharing risks and benefits, long term perspective on products, skills, and impact) as well as a 'model' client (compensate downturns, first user, demonstrator). A model client is not only a buyer, but also a partner interested in developing a close relationship and supporting sustainable development of the suppliers as well as respecting community values. This role is not really new and has had existed in some countries, particularly in France, for many years. However, we may see an emerging change in the type of relationship where much more shared values and mutual trust are being developed;
- Developing 'participative' standards for protecting and promoting shared social values (safety, environment and other social values), elaborated through a consultative process that supports the creation and evaluation of new standards, as well as with the help of effective mechanisms for establishing priorities. This role is evolving towards a more participative approach, where decision-making is based on expanded consultations and control is being decentralized, although coordinated. The process is becoming complex with the increased participation of different stakeholders (manufacturers, designers, trade and general contractors, and representatives from users). At the same time, the state and the industry must maintain its leadership in developing new international standards and/or in international harmonization of current ones.

Although these types of interventions are common in many countries, again most probably because of increasing international shared values and beliefs, specific actions will need to be developed in each country in order to provide a niche position and competitive advantage on international market. It does not mean to close our borders, but rather develop a unique expertise that will be complementary to others and desired. A common basic knowledge and standards are needed for facilitating global exchanges; however, unique or rare competencies are required for maintaining a competitive advantage.

Specific public instruments or actions for supporting innovation are far from being obvious to define. The industry must find a good return on investment (ROI) from innovation, and this has not been so easy thus far. A number of challenges are facing policy makers. What is the best way to reach SMEs? How will construction firms capture benefits from innovation? Beyond technology, a better understanding of the business – competition, culture, and customers – is required (Seaden, 2002).

The three roles described above have to work together, in a connected manner, learning from each other. They must be aligned on developing a strategic position in international markets by developing niche sectors (through negotiation and coalition), and by establishing new standards in these sectors. Although international partnering and trade are still relatively marginal in construction, global knowledge flows, best practices, standards and technologies circulate very rapidly. Moreover, highly profitable markets are already attracting international competition. Firms must be aware of international development for maintaining their market.

References

Amable, A., Barre, R. and Boyer, R. (1997), *Les systèmes d'innovation à l'ère de la globalization*, Ed. Economica, Paris.
Blyer, M. and Coff, R. W. (2003), 'Dynamic Capabilities, Social Capital, and Rent Appropriation: Ties That Split Pies', *Strategic Management Journal*, vol. 24, pp. 677-686.
Campagnac, E. (1998), 'National system of innovation in France: Plan Construction et Architecture', *Building Research and Information.* Vol. 26, no. 5, pp. 297-302.
Campagnac, E. (2000), 'The contracting system in the French construction industry: actors and institutions', *Building Research and Information*, vol. 28, no. 2, pp. 131-140.
Campagnac, E. (2001), 'La 'commande' comme nouveau marché de services : crise ou renouveau du professionalisme? Les leçons de l'expérience britanique', *Espaces et Sociétés, Projet urbain, maîtrise d'ouvrage, commande*, vol. 105-106, no. 2-3, pp. 17-55.
Construction Research and Innovation Strategy Panel (CRISP) (1997), 'Creating a Climate of Innovation in Construction', CRISP Motivation Group, London, UK. Working document.
CRIC (Centre for Research on Innovation & Competition – University of Manchester). (2002), 'Innovation in the Service Sector – Analysis of data collected under the Community Innovation Survey (CIS-2)', CRIC Working Paper No. 11.
Dasgupta, P. (2002), 'Social Capital and Economic Performance: Analytics', Working Paper, University of Cambridge, Faculty of Economics.
Fairclough, Sir John (2002), 'Rethinking Construction Innovation and Research: A Review of Government R&D Policies and Practices', Department of Trade and Industry & Department of Transport, Local Government Region, London.
Foray, D. (1993), 'Standardization et concurrence: des relations ambivalentes', *Revue d'économie industrielle*, no. 63, Premier trimestre, pp. 84-101.
Frinking, E., Hjelt, M., Essers, I., Luoma, P. and Mahroum, S., (2002), *Benchmarking Innovation Systems: Government Funding for R&D*, Technology Review no. 122/2002. Tekes - National Technology Agency, Espoo, Finland.
Gann, D.M. (1998), 'Learning and Innovation Management in Project Based Firms', 2nd International Conference on Technology Policy and Innovation, August 1998, Lisbon, Portugal.
Kuttner, R. (1997), 'Whistling past the graveyard in Asia', *Business Week*, no. 3557, pp. 26ff.
Manseau, A. and Seaden, G. Eds. (2001), *Innovation in Construction – An International Review of Public Policies*, Spon Press, London.
National Research Council of Canada (NRCC) and the International Council for Research and Innovation in Building and Construction (CIB) (2001), 'Construction Innovation: Opportunities for Better Value and Profitability', Symposium, Ottawa (Canada), June 6-15, 2001.
NSCT – National Science and Technology Council (1999), 'Construction Building: Federal Research and Development in Support of the U.S. Construction Industry', U.S. Federal Government, Washington.

OECD (1996), *The Knowledge –based Economy*, OECD, Paris.
OECD (1997a), 'Proposed Guidelines for Collecting and Interpreting Technological Innovation Data', (Oslo Manual), OECD, Paris.
OECD (1997b), 'National Innovation Systems', OECD, Paris.
OECD (1998a), 'Science, Technology and Industry Outlook-Construction', OECD, Paris.
Porter, M. (1998), *On Competition*, Harvard Business Review Book Series, Cambridge, USA.
Portes, A. (1998), 'Social capital: its Origins and Applications in Modern Sociology', *Annual Review of Sociology*, vol. 24, pp. 1-24.
PRI-SSHRC (Policy Research Initiative – funded by the Social Science and Humanities Research Council) (2003), 'Social Capital: Building on a Network-Based Approach' Discussion Paper.
Saviotti, P. P. (1995), 'Renouveau des politiques industrielles: Le point de vue des théories évolutionnistes', *Revue d'économie industrielle*, no. 71, pp. 199-212.
Seaden, G. (1996), 'Economics of Technology Development for the Construction Industry', CIB Report Publication, no. 202, Rotterdam, Netherlands.
Seaden, G. (2002), 'Changing more than R&D: responding to the Fairclough Review', *Building Research & Information*, vol. 30, no. 5, pp. 312-315.
Winch, G. (1996), 'The Contracting System in British Construction: The Rigidities of Flexibility', Le Groupe Bagnolet Working Paper No. 6, University College London, London, UK.
Winch, G. (1998), 'Zephyrs of Creative Destruction: Understanding the Management of Innovation in Construction', *Building Research & Information*, vol. 26, no.4, pp. 268-279.
Winch, G. (2000), 'Institutional Reform in British Construction: Partnering and Private Finance', *Building Research and Information*, vol. 28, no. 1, pp. 141-155.
Winch, G. and Campagnac, E. (1995), 'The Organization of Building Projects: An Anglo/French Comparison', *Construction Management and Economics*, vol. 13, no.1, pp. 3-15.

Chapter 8

Construction Industry Paradigms: The Final Frontier

Andre Dorée and Frens Pries

8.1 Need for Change; Resistance to Change

For many decades the construction industry has remained a staid, tradition-bound sector due to its relatively stable environment. Recently the environment of the construction industry has changed and it is now generally agreed that there is a need for more innovation and change in the construction industry (Slaughter 1998, Winch 1998). This implies that enterprises will have to operate in a more 'market-driven' manner. Management will have to link 'technology' to 'market'. However, several authors point out that the construction industry is not a rapidly changing sector that smoothly adopts innovations. Manseau (1998) argues that the sector appears to change slowly and with great difficulty. In other chapters of this book this reluctance to change has also been addressed.

The huge number of studies and papers produced by researchers analyzing the construction industry suggests that major changes are at hand; in practice, however, nothing really changes. The construction sector seems impotent to change. The causes for low rates of innovation have been investigated, and among other issues, management peculiarities of construction have been pointed out as explanations.

8.2 Intermezzo: Is the Construction Industry Really Extraordinary?

In Europe the construction industry is known as a mature, traditional industry. Nam and Tatum (1988) argued that the characteristics of the constructed products result in limitations for construction technology. These views are based on observations on lagging productivity, poor quality and high costs of construction in practically all industrialized nations:

- Although various different construction processes are distinguished, the traditional organization of the building process is a core item in most

studies (e.g. Bakens 1992, Koskela and Vrijhoef 2001): 'In all Western industrialized countries people recognize or are starting to recognize the traditional, segmented organization of the building process as a major problem in general and as a major hindrance for innovation in particular';
- The Government has an especially dominant influence on this market (Pries 1995). Due to technical regulations, the quality of a major part of the production is determined according to strict regulations. These technical regulations and licenses, both national and local, have an especially dominant influence on social housing, as well as in civil engineering. This leads to limited opportunities for product variation. What remains is a very severe price-competition;
- The products (buildings, bridges, roads, etc) can be clearly distinguished. They are location-bound, have a very long life span and high costs. Furthermore, the long life span (on average 100 years for a building) compels customers to stick to proven methods (and avoid radical changes); they have to live with it for many years;
- Furthermore, the construction sector can be characterized by the great number of small enterprises and varying collaborations; 'co-makership' (or other strategic alliances) hardly exists. The emphasis lays on operational (project) management. For most firms, strategic management does not exist (Bakens, 1988): 'commonly the horizon of contractors is not beyond the moment of completion of a project'. Finally, the weak patent possibilities, low barriers of entry and the poor image of the industry must be mentioned. In combination, these characteristics have a great effect on the innovative behavior of the sector.

These peculiarities of the industry are often used as an explanation for its behavior in general and more specifically, for the lack of innovation and change within the industry. This may partly be the case, but a comparison with other industrial sectors shows that some share the same characteristics. Agriculture is very much location-bound, products like high-speed trains, ships and rockets also have high initial and exploitation costs and severe price-competition. Sectors such as the pharmaceutical and nuclear industry face severe governmental influence and regulations, severe price competition and have a great influence on quality of life. Every single one of these sectors has experienced major changes. They have faced concentration, paradigm changes (for example the automotive industry moving from a technologically-driven towards a market-driven sector), major ICT-investments, implementation of supply chain management, 'co-makership' and early supplier involvement, etc. The construction industry shares all of these characteristics and faces similar challenges.

The debate on the construction industry's characteristics aside, the question is relevant anyhow. Why is it that other sectors seem to change more easily than the construction industry? In our opinion, it is not enough to merely point out the specific characteristics of the sector. Also causes and effects may be confused. Traditional processes, the lack of real cooperation and the high degree of fragmentation may not cause the specific behavior of the sector; they might very well be the effect of the sectoral behavior. The management paradigm of the industry is also one explanation for the specific characteristics of the industry.

8.3 The Innovative Behavior of the Construction Industry

There are a variety of business drivers that point to the need for the construction industry to put a stronger focus on innovation. A parade of documents illustrates the growing recent attention to this subject (Slaughter 1998, Winch 1998, Atkin 1999, Gann 1998 and 2000, Koskela and Vrijhoef 2001). A quick consultation in the ARCOM abstract database (www.arcom.ac.uk), with 'innovation' as keyword for example, produces 26 hits in the last five years and only four hits in the years before that. However, when it comes to innovation, the construction industry cannot boast a reputation as strong as other industries.

Quantitative data on innovation in the construction industry are rare. Often, discussion is based on a restricted number of cases and statistical material on the percentage of turnover spent on R and D. Where construction spends 0.3 per cent of turn-over on R and D activities, the pharmaceutical industry may spend almost ten per cent (Pries, 1995)! Again, our industry looks comparatively weak, but these figures are not really comparable due to variations in underlying definitions. Agencies often use the Frascati-manual of the OECD to define R and D-efforts (see also Chapter 4). Within this definition, applied research is only counted when used to produce new or substantially improved materials or processes (Pries, 1995). The adjective 'substantial' is crucial in this case because it excludes incremental innovation. The overwhelming majority of all innovations in the construction industry are incremental. Frascati-based measurements discriminate against the construction industry by under-reporting innovation. This study provides quantitative insight into the innovative performance of the Dutch construction industry. The method applied is described in the next section, followed by a presentation of the results and our suggestions for interpretation.

8.3.1 Methods

The main objective is of the study was to track the changes in the level of innovation in the Dutch construction industry. In this study, we posed the following question: what type of innovations dominated the last fifty years? Who were the main innovators? Finally, how did the innovative behavior vary in time? A database was compiled to perform statistical routines and analysis. The data on innovations were collected in two ways:

- Literature search: renewal in the Dutch construction industry is fairly well documented up until World War II, mainly by Priemus (1970);
- An analysis of 55 years of publications in two leading Dutch professional journals (BOUW and Bouwwereld). Using professional journals as a source causes a methodological problem. The journals portray the industry. When editorial policies change, this portrait alters; not because the industry changes, but because it is projected differently. Shifts in the data might therefore be the result of publishers' interpretation, rather than an actual industry change. These journals, however, appear to have been consistent on policy and subscribers.

The characteristics of every innovation mentioned were recorded (type of innovator, year, sort of innovation, etc.). A database of 492 innovations introduced by Dutch companies from 1901 to 2000 was created. The analysis is restricted to residential and non-residential building to reduce heterogeneity. Such a dataset is typical for the Dutch construction industry. Nevertheless, the findings seem to match international developments, which we will address this in the discussion.

8.3.2 Results

The majority of innovation originates from suppliers. Out of 492 relevant innovations, 70 could not be attributed to a particular party (Figure 1). Of the other 422 items, suppliers (within the construction industry and external to it) proved to be the main source of innovation, producing 65 percent of all registered innovations- just over half of the process innovations and almost 80 percent of product innovations. The contractors' role in innovation, although limited and mainly restricted to process innovation, seems to be on the rise.

	Innovation: All	Process	Product
Contractor	10,9%	18,2%	3,4%
Supplier	64,6%	50,9%	78,4%
Arch./consultant	8,8%	10,7%	7,2%
Miscellaneous	15,7%	20,1%	11,1%
n =	422	214	208

Figure 8.1 Parties and types of innovations

Innovations Come from Different Sources: The sources of innovations were categorized by industry sector. Because observations were made from the point of view of the construction industry, it is logical that most of the documented innovations (63 per cent) originate from that sector. However, the influence of other industries is enormous (Figure 2). About 40 per cent of all innovations from other sectors originate from the chemical industry. The metal industry, electrical engineering and machinery also play an important role.

The construction industry, then, is strongly dependent on other industries. Seaden (2001) states that firms have a choice of having their future shaped by processes and technologies developed elsewhere (e.g. equipment and material suppliers) or they can be 'market makers'. Our analysis does not provide the ultimate answer to this question, but it seems defensible to respond that innovation in the construction industry in the Netherlands – and probably other European countries – is highly dependent on other industries and that there are few 'market makers'.

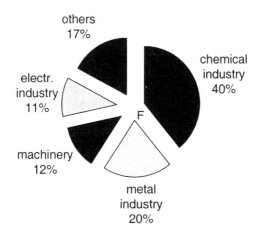

Figure 8.2 Sources of innovation

Innovation Seems to be on the Rise: Several unique periods can be distinguished in the history of innovation in the construction industry. After 1945, the post-war reconstruction in the Netherlands led to the development of various ('industrial') building systems. Starting in 1964, large-scale concrete pre-cast systems became very popular. These systems achieved a market share of over 50 per cent in just a few years. Because of changing production (more variation and on average smaller projects) starting in the mid-seventies, small-scale stacking elements (blocks and bricks) became the most popular building systems.

Recently, construction output in the Netherlands has become more varied and projects have become smaller in size. As the housing-market slowed down (prices reached a maximum level, in some Dutch regions a demand-market emerged) many construction companies and especially property developers invested in new consumer-oriented housing-concepts. This has contributed to a serious increase introduced per year by construction companies of the number of innovations (Figure 3).

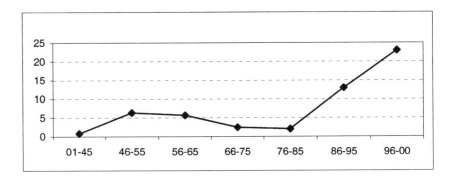

Figure 8.3 Average number of innovations per year

The Majority of Innovation is Incremental: In the professional journals studied, the relative frequency of incremental (by far the most frequent kind of innovation) compared with radical innovations (often consisting of 'families' of incremental innovations) does not vary much over the time period monitored. Per period, the number of incremental innovations can be five to nine times higher than radical innovations. As radical innovations are often more explicitly documented, it is likely that these figures are even more extreme in reality.

Innovation as an Art of Cooperation: The degree of collaboration in the innovation process was defined as a parameter. Apart from individual enterprises (when explicitly stated in the sources), collaborations of two or more enterprises and collective R and D (sector, national and international) were documented. There were no examples of international activity in our sources. The analysis also shows that most of the reported innovations are the result of enterprises operating individually. From the beginning of the eighties up until the present day, cooperation has become popular; 50 per cent of all innovative-activities are the result of some kind of collaboration. This cooperation is mainly to be characterized as cooperation between two or more firms on the one hand, and (sub) sector-activities on the other hand. National and international partnerships and programs play a very modest role. An image emerges of innovation activities taking place within one firm, or between a small set of enterprises. Collective programs play a modest role. This is in line with the findings of Seaden and Manseau (2001) who state that most of the currently available public policy instruments in support of innovation have not been of great use to the construction industry.

Size of Enterprises does Matter, But Not Much: Are smaller firms better innovators? This is a classic question. Schumpeter first believed large companies could not innovate (Schumpeter mark I). Later he changed this view (mark II – see also Chapter 1; Martin and Scott 2000). The EU innovation policies strongly focus on small and medium size firms. What does our data tell us? Innovations were analyzed based on the size of the enterprises involved. In most periods, the majority of the innovations emerged in the smaller enterprises (about 60-70 per cent). Hence, smaller enterprises play an important role.

On the other hand, it cannot be concluded that small companies play a dominant role, because about 88 per cent of all construction companies in the Dutch construction sector have less than 10 employees (CBS, 2002). The biggest firms only represent a smaller percentage of the total construction market. These bigger firms (over 100 employees) are often minimally integrated conglomerates of SMEs. This further speaks to the notion that there are few, if any, very large individual construction firms in the Netherlands. It is particularly interesting to note that smaller enterprises tend to be more involved in process-innovation while the larger firms have a stronger track record in product-innovation. This makes sense as process innovation (new equipment and organizational on-site renewal) in construction has more small-scale characteristics than product innovation.

Regulations Spark Innovation: MacMillan (2001) points to the central role that governments play in supporting innovation via the regulatory framework. Our analysis underscores the important role of government regulation. Innovations are labeled in relation to changes in Dutch national building regulations or building codes. This demonstrates that over 30 per cent of all innovations are the result of new regulations. Until 1975, just over 20 percent of registered innovations were related to new regulations. Since 1975 this has risen to almost 40 per cent of registered items. The regulations concerning safety and environmental impact have been particularly dominant in the last two decades, but labor conditions remain an important factor. Gann (1988) looks at different ways to draft regulation. They make a case for a more flexible performance based form of standards. Such an institutional framework allows firms to innovate within a competitive context.

The Market Needs to be Discovered: The data show that incremental innovation (a small change with limited impact on surrounding elements –see also Chapter 1) is by far the most important innovation from a numerical perspective. When we consider the relatively hidden nature of this kind of innovation, in reality there should be an even greater importance of these small innovative steps (Pries, 1995). What drove these innovators? Throughout the period studied, the primary motive for innovation was to improve productivity (75 per cent). Only 25 per cent of innovations were in response to specific market demands. Innovation in the construction sector is most often a function of productivity considerations. Although in the last period considered the market motive is growing, it is still relatively unimportant. The construction industry continues to be inward looking towards technology, and strategy is rarely applied to translate customer needs into more innovative and differentiating propositions.

8.4 Reflections and Questions

Innovation is an important topic for the construction industry. No matter how strong or weak its performance is compared to other industries, no matter how strong or weak its absolute performance – if we could find a measure for it, and agree upon it – changing the performance of the construction industry depends on the ability to innovate. Therefore, we need further research on how the industry deals with innovation.

Innovation in the construction industry is mainly incremental, most likely due to the fact that in general, firms are more inward looking with regards to improving their technology and related processes. The market is perceived as price and cost driven. Many small and medium sized firms produce similar products with similar technology and materials. Their focus is mainly on projects and project control (Gann and Salter 2000). Business continuity is understood as dependent upon securing a stable workload and efficient utilization of resources in addition to lowering operational cost.

Suppliers have higher up front investments in technology. Given the risks of losing these investments, they feel compelled to 'create' their market niches and subsequently there is a stronger need to innovate. The imperative to cooperate on innovation might be an indirect function of mean, lean and flexible strategies. The complexity of products requires increased coordination along the supply chain. New regulations, for product as well as production, change the business environment. Firms have to adjust in order to avoid going bust. An increasing number of competitive firms see these changes as business opportunities.

In several countries, industry and government initiatives have been taken to awaken the industry out of its inward looking, traditional patterns. For most firms, it is going to be of paramount importance to understand demand and supply drivers in their business environment. Their challenge is to connect an understanding of their markets, ever-changing regulations, and the technology of their suppliers and partners. The more pro-active the firm, the greater the chance of it achieving a sustainable competitive advantage. Such firms will have an opportunity to lead the industry to a higher performance standard. Will this happen? How can the presumed lack of sectoral change be explained and how can we deal with this? We will first introduce the notion of a *paradigm* as a concept, in the next section. Then, as part of the explanation of slow change, go deeper into the management paradigm in construction industry.

8.5 Paradigms

The acknowledgement of social dynamics and behavioral perspectives on change processes inspired researchers in the past. In 1935, Gilfillan analyzed innovation in shipbuilding and found that an invention is very much a new combination from the 'prior art'. Gareth Morgan sees the work of Kuhn on paradigms and paradigm shifts as an important managerial concept (in Morgan 1996). The paradigm concept was introduced by Kuhn (1962)in the context of scientific development. Currently, the concept is used in other contexts as well, and has made its way to organization theory and the contemporary literature on innovation management and strategic management (Morgan 1996).

A paradigm can be described as the system of knowledge, values, and assumptions that guide observations, decisions and actions. It acts as a pair of glasses or a filter that helps to make sense of reality. A human being acquires his or her paradigm through education, experience and by socializing with groups of which he or she is part (Koskela, 1992). In a certain way, it is necessary for members in a group to hold the same paradigm. It simplifies the mutual communication considerably when the group members have a similar 'frame of mind'. Koskela (1992) also recognizes the implicitness of paradigms: a paradigm is usually adopted unnoticed, and is seen as a part of a 'practitioner's knowledge'. Quite often the 'owners' of a paradigm take it for granted that their paradigm is the only reasonable way to look at the world. Therefore, it is difficult to discuss or break a paradigm.

Various authors in contemporary management literature have used Kuhn's concept of paradigms and paradigm shifts in their research as well (see also Morgan 1996). Related models in innovation-management are those of Nelson-Winter and Dosi, who applied the ideas of Kuhn to technological development. In these theories innovation is seen as the result of an interaction between 'technological paradigms' on one side and a 'selection environment' on the other. A technological paradigm was defined by Dosi (1982) as a 'model' and a 'pattern' of solution of selected technological problems, based on selected principles derived from natural sciences and on selected material technologies. A technological paradigm embodies strong prescriptions on the directions of technical change to pursue and to neglect.

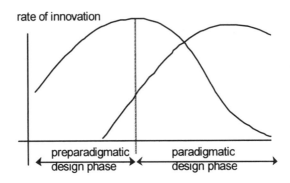

Figure 8.4 Stages in the innovation process

Enterprises operate within these technological paradigms, thereby predetermining the direction of R and D and technological solutions. An innovation fitting in a certain paradigm is adopted more swiftly than complete new or radical innovations. Several authors worked along the line of these technological paradigms and theories of path dependence. Gardiner (1986) uses the term trajectory. Nelson-Winter (1982) analyze technological progress in terms of 'technological regimes.' The term paradigm also emerges in the work of Teece (1989), who distinguishes two stages in an innovation process: the preparadigmatical stage and the paradigmatical stage (Figure 8.4, see also Abernathy and Utterback 1988).

These economically-oriented theories are further developed in theories of the social construction of technology. This research shows that it is not the level or type of technological knowledge that is decisive, but the relevant social groups that are in the position to exercise decision-making power. Barker (1993) connects paradigm shifts and innovation. To innovate, people and organizations need to be able to shift their paradigms. In order to develop and to adopt a new technology or managerial concept, it is crucial to have a notion that something 'new' (better, cheaper, faster, etc.) can be done with the new discovery. However, the judgement of the new discovery initially takes place based on ruling, established convictions.

The paradigm concept is discussed at different scales: paradigms exist at the level of individuals, of organizations and of industries, communities (Kuhn 1962) and nations. At this point, we have no real clear understanding how paradigms at these different levels interact. To some extent, the notion of the importance of leaders (as individuals) suggests that these levels do add up, in the sense that organizational paradigms are the outcome of individual insight, bias and leadership. But there is not only a unidirectional influence of individual paradigms on the organizational paradigm: individuals shape their paradigms in a complex process of socialization and in the interaction with other individuals and groups. Argyris (1993) mentions the concept of 'defensive routines'; people, and therefore also managers, tend to avoid subjects which go against the grain, which leads to a reluctance to learn. Natural selection reinforces the paradigm. The ultimate effect of this is 'group think'.

Paradigms are thus an acknowledged phenomenon. Paradigms tie social, scientific and economic communities together and create trajectories to the future, building on the past. However, paradigms also have a conserving property, which may hinder change. Since these paradigms have a strong tacit foundation, the actors are often unaware of the implicit guidance by the paradigm. Paradigms may obstruct change processes, but as long as the underlying preferences are unknown, they cannot be dealt with directly.

8.6 The Management Paradigm in the Construction Industry

Top management is responsible for the strategic direction of an organization. The managers are selected and appointed on the basis of their capability to address these issues. They need to read the changes in the business environment and translate them into internal policies and policy deployment. Management paradigms refer to implicit and explicit ideas and notions of good management and good managers; to the type of management and manager that are required to make an enterprise successful. These management paradigms are typically related to business objectives given the environment, and vary per type of organization and type of business. Since management paradigms have an impact on change processes that the construction industry is undergoing, management paradigms are a relevant issue to consider. It might explain why change processes develop as they do, and might also help to deduce improvements in the change initiatives. So, what is the dominant paradigm in construction industry?

In 1995, Pries and Janszen described the dominant paradigm in construction as the 'engineer's paradigm'. They dug into the qualifications of the managers of the 100 largest construction companies in the Netherlands. The construction sector exceeds other sectors in the technical education and project background held by management. These statistics, however, do not focus on managerial functions in the manufacturing industry where, for example, 69 per cent of all positions are technical professions, while in the construction industry 81 per cent are technical professions.

Management in construction can be characterized by 'the lack of orientation on [toward] the future, the lack of strategic awareness' (Bakens, 1988). This is stipulated in several publications (Pries, 1995, Egan 1998). No career is possible without extensive site experience. Obtaining management capabilities requires training on the job and learning the twists and turns of the ropes from one's superiors.

8.6.1 Technical Certification and Employment Patterns

As stated before the construction industry has faced an increasingly turbulent environment. These changes are likely to have effect on the people and organizations within the sector. When the environment changes new competencies are needed. In order to find out if new competencies are required, the educational degrees of the executive management of Dutch construction companies in 1992 and 2002 were compared (Source: Registers of the Chamber of Commerce and Internet version of ABC bedrijfs-en productinformatie). In 2002, the profiles of a total of 2082 construction executive managers were recorded and analyzed. For every person we recorded the following:

- the size of the company (number of employees);
- title of the executive management, with a division of: no title/not mentioned, ir (Dutch equivalent of M.Sc.), ing. (Dutch equivalent of technical B.Sc.), drs. (Dutch equivalent of MBA, ME), B.Sc., M.Sc., Prof. (professor) and others.

A comparison of construction management employment advertisements in 1982 and 2002 was carried out (both September and October for each year as this is the most turbulent period in the labor market). In employment advertisements, required skills are concisely formulated. In Holland *Intermediair* is a leading magazine for employment advertisements for the higher segment of the labor market. A total of 1078 advertisements (all sectors) were analyzed (140 from the construction industry).

Our conclusions were very cautiously interpreted because approach has methodological limitations. Hence:

- Listing a title or qualification can be characteristic of a certain era. It is possible that people have titles, but that they are not mentioned in the register;
- No distinction was possible between 'not mentioned' and 'no title';
- The contents of educational programs have changed over time. It is methodologically incorrect to gratuitously compare diplomas from 1970 with a diploma from 1998;
- Interpretation of the classification in disciplines and functions (an educated guess was required in some instances);
- In the early eighties there was an economic relapse in Holland;
- Advertisements are only a part of the recruitment and HRM process in companies. We have no additional data on internal courses, also leading to competency shifts.

8.6.2 Results: Construction Management Employment Advertisements

A total of 1078 advertisements were found. The separation between construction and other sectors is stable when 1981 is compared with 2001 (construction accounts for about 12–13 per cent of all advertisements). The total number of advertisements in 2001 has grown compared with 1981 (Figure 5). This may have several causes, of which the economical situation will probably be the most important.

A division in disciplines shows that technical disciplines are by far dominant. Over 60 per cent of all advertisements focus on technical knowledge, skills and experience (Figure 6). Furthermore, general management competencies are needed; however, in most advertisements concerning general management, the call for practical (project-) experience is strong.

Distribution of Advertisements	1981	2001
Construction	11,9%	13,7%
Other sectors	88,1%	86,3%
Total	100,0%	100,0%
	N=452	N=627

Figure 8.5 Distribution of advertisements: Construction and other sectors

Distribution of disciplines (construction only)	1981	2001
Technical	61,1%	64,0%
Legal	0,0%	5,8%
Management	11,1%	11,6%
HRM	3,7%	3,5%
Financial	7,4%	4,7%
Various	16,7%	10,5%
Total	100,0%	100,0%
	n=54	n=86

Figure 8.6 Distribution of disciplines

Striking is the growth of positions requiring legal qualifications. In 1981, zero advertisements were recorded and almost 6 per cent in 2001 (Figure 6). Apparently, there is a growing need for contractual and legal competencies.

The dominance of the technical component is striking when compared with non-construction sectors. Across all sectors, only 24 per cent of all advertisements required technical competencies in 1981. In 2001, this was reduced to a mere 6 per cent. Obviously in the construction industry (we did not investigate whether other single sectors have the same peculiarity) there is a constant and exceptionally high need for technical skills and experiences. In other sectors, the general need for technical skills and experiences has been reduced over the last 20 years (Figure 7).

Distribution of disciplines	1981		2001	
	Construction	Other sectors	Construction	Other sectors
Technical	61%	24%	64%	6%
Non-technical	39%	76%	36%	94%
Total	100%	100%	100%	100%
	n=54	n=392	n=86	n=342

Figure 8.7 Distribution of advertisements

8.6.3 Results: Education Level of the Management

In 1993 an analysis was made of the larger construction companies (over 100 employees) surveyed titles held by general managers. In this comprehensive analysis, 53 companies were analyzed. About 51 per cent of these managers had a technical title (Ir. or Ing. are equivalents of MSc or BSc degrees), two per cent had a legal education, four per cent MBA and 43 per cent did not bear a title (Pries, 1995).

	>100	all
	2002	2002
No title	71%	83%
Title, technical	24%	14%
Title, non-technical	5%	2%
No title & title technical	95%	98%
	n=729	n=2082

Figure 8.8 Distribution of advertisements by qualification

It can be expected that for the people without a title, the overwhelming majority originate from construction practice. They have advanced into general management positions. Of all recorded titles, 90 per cent were engineers.

For 2002 a total of 2082 executive managers were recorded in our database. 83 per cent of all general managers in the construction sector do not bear a title (Figure 8). Of those who bear a title (14 per cent of all), 83 per cent are engineers.

Non-technical titles are rare with general managers in construction who hold a university degree. This is even more the case with smaller enterprises (Figure 9).

Company size (nr. of employees)	No title	Ir./Ing.	Other
1 to 10	95%	4%	1%
11 to 50	93%	7%	1%
51 to 100	80%	17%	3%
101 to 250	79%	16%	5%
251 to 500	63%	34%	3%
501 to 1000	58%	32%	10%
> 1000	58%	38%	5%
Total	83%	14%	2%
n =2082			

Figure 8.9 Distribution of advertisements by firm size

8.6.4 Conclusions on Managers' Education

The most sought after competencies of managers in the construction industry does not vary much over time. There is a constant and dominant pursuit of technical people. The dominance of the technical component (over 60 per cent of all construction advertisements) is particularly striking when compared with other sectors. For all other, only 24 per cent of all advertisements required technical competencies in 1981. In 2001 this was reduced to a mere six per cent. The construction industry does not follow that trend.

Although we might argue the business environment changed for the construction industry also, and that subsequently new approaches to management are required, the general managers in the construction industry are still predominately technically trained people. 95-98 per cent of all general managers bear a technical title or are promoted through the ranks based on their experience in project-practice. Of those who bear a title, 83 per cent are engineers. Non-technical degrees are rare among general managers in construction. Furthermore, we found that: the smaller the enterprise, the smaller the number of titles within general management; and the smaller the enterprise, the fewer managers without a technical education. These findings seem to be rather constant over time.

Both support the argument that the core competencies of general managers in construction are technical. We can conclude that, in spite of a growing need for more strategic marketing skills, the engineer's paradigm does not change at all! Apparently, the ruling (engineer's) management paradigm is very strongly rooted in the sector.

8.7 Traditionalism and the Industry Paradigm: Our Final Frontier?

Other sectors, such as the textile, shipbuilding, or automotive industries have thought for years that their sectors would not change. At first, change and innovation were just topics for symposia, consultants and journals, and not subjects for action. These industries changed their management paradigms partly through crises. The same change discussions can now be found in the construction industry. In our opinion, it is not a question of whether or not this sector will change, but when these changes will take place. Heraclites stated more than two thousand years ago: 'nothing is as permanent as change'. How can we deal with this typical sector, using this paradigm? Is there a model for change? Pries (1995) provides an overview of models for sectoral change:

- Consensus model. Together parties in the construction industry decide to change. This model has been generally adopted in the last 30 years, and it has not worked for the 30 years. The degree of fragmentation in the sector and the lack of mutual trust may have caused this. Even now, both the Dutch government and interest groups tend to embrace the consensus model;
- Top-down model (the iron fist), with as variants:
    - Dominant government: this centralistic approach does not exist in Western Europe;
    - Dominant enterprises: a restricted number of large dominant companies have the power to set certain standards. In the European construction market this situation does not exist, although in certain construction sub-markets a certain degree of market-dominance (or market-organisation) seems to exist;
    - Dominant buyers: a restricted number of large buyer dominate the sector. In most countries, the government could be such a dominant party, but in practice this innovation-stimulating role is very much capable of improvement. The model of dominant buyers exists in theory, but does not function in practice;
- Pioneers-model. An enterprise or a number of enterprises are successful innovators. In this model, open systems do not exist; enterprises work in closed systems. When successful the competitors will follow swiftly.

We believe that change in the short term will only occur because of the pioneers-model. In every sector and in every era there are enterprises that are more successful than others are. Of all the models, the pioneers are most likely to change the construction sector. In the long-term, the top-down model will become increasingly more important should the government adopt more open acquisition policies that promote long-term value and performance (rather than initial costs). This will definitely stimulate innovation and change in the sector (Seaden and Manseau, 2001). Besides government, the role of the consumer or client will become increasingly important. When consumers unite and when instruments (such as visualization) democratize the construction process, the consumer could at last become a major player.

As we have seen, the construction company of the future is in need of new managerial capabilities. Management is currently dominated by engineers. Their competencies have to change from technological and introverted towards extroverted and market orientated. An important role is reserved for education. When Scherer (1992) concludes that: 'it is essential that management schools provide their students with in-depth education on the significance of technological innovation, the challenges it poses, and means of sustaining it', it can be suggested that the opposite is necessary for technical (construction) education.

References

Abernathy, W.J. and Utterback, J.M. (1988), 'Patterns of Industrial Innovation, Technology Review. In *Strategic Management of Technology and Innovation*, Burgelman et.al (eds), Irwin, New York. pp.141 - 148.
Argyris C. (1993), *On Organizational Learning*, Blackwell, Cambridge (Mass.).
Argyris C. and Schön, D. A. (1978), *Organizational Learning; a Theory of Action Perspective*, Addison-Wesley, Reading (Mass.).
Atkin B. (1999), *Innovation in the Construction Sector*; ECCREDI.
Bakens, W.J.P. (1988), *Bouwen aan 2005; toekomstperspectief voor managers, bestuurders en onderzoekers in de bouw*, Tutein Nolthenius, Amsterdam.
Barker J. (1993), *Paradigms: the Business of Discovering the Future*, HarperBusiness, New York.
Dosi, G. (1982), 'Technological Paradigms and Technological Trajectories', *Research Policy*, vol. 11. pp.147-162.
Egan, J. (1998), *Rethinking Construction*, Department of the Environment, Transport and the Regions, London.
Gann D. and Salter A.J. (2000), 'Innovation in Project-based, Service-enhanced Firms: The Construction of Complex Products and Systems', *Research Policy*, vol. 29, pp. 955-72.
Gann, D.M., Wang, Y., and Hawkins, R. (1998), 'Do Regulations Encourage Innovation?: The Case of Energy Efficiency in Housing', *Building Research and Information*, vol. 26, no. 4, pp. 280–296.
Gardiner, J.P., (n.d.), 'Design Trajectories for Airplanes and Automobiles during the past Fifty Years' in C. Freeman (ed). *Design, Innovation and Long Cycles in Economic Development*, Francis Pinter, London. pp. 121 - 142.
Koskela, L. (1992), 'The Application of the New Production Philosophy to Construction', *CIFE Technical Report*, no. 72, Stanford University, Palo Alto Cal.
Koskela, L., and Vrijhoef, R. (2001), 'Is the Current Theory of Construction a Hindrance to Innovation?' *Building Research and Information*, vol. 29, no. 3, pp. 197-207.
Kuhn, (1962), *The Structure of Scientific Revolutions*, Chicago: University of Chicago Press.
Kululanga, G.K., Edum-Fotwe, F.T., and McCaffer, R. (2001), 'Measuring Construction Contractors' Organizational Learning', *Building Research and Information*, vol. 29 no. 1, pp. 21-29.

Latham (1994), *Constructing the Team*, HMSO, London Report.

MacMillan, S. (2001), 'How does Innovation Link to Research?' *Building Research and Information*, vol. 29, no. 3, pp. 250-252.

Manseau, A. (1998), 'Who Cares about Overall Industry Innovativeness?' *Building Research and Information*, vol. 26, no. 4, pp. 241–245.

Martin S. and Scott J.T. (2000), 'The Nature of Innovation Market Failure and the Design of Public Support for Private Innovation', *Research Policy*, vol. 29, pp. 437-47.

Morgan (1996), *Images of Organization*, 2nd edition, Sage Publications.

Nelson, R.R. and S.G. Winter (1977), 'In Search of a Useful Theory of Innovation', *Research Policy*, vol. 6, no. 1, pp.36-76.

Priemus, H. and R.S.F.J. van Elk (1970), 'Niet-traditionele woningbouwmethoden in Nederland, *Stichting Bouwresearch*, no. 26.

Pries, F. (1995), *Innovatie in de bouwnijverheid* (in Dutch: Innovation in the construction industry), Eburon, Delft.

Pries, F. and F. Janszen, (1995), 'Innovation in the Construction Industry: The Dominant Role of the Environment', *Construction Management and Economics*, vol. 13, no. 1, pp. 43-51.

Scherer, F.M. (1992), *International High-Technology Competition*, Harvard University Press, Cambridge.

Seaden, G., and Manseau, A. (2001), 'Public Policy and Construction Innovation', *Building Research and Information*, vol. 29, no. 3, pp. 182-196.

Seaden, G.. (2001), 'Changing Technology or Managing Change?' *Building Research and Information*, vol. 29, no. 3, pp. 248-249.

Slaughter, E.S. (1998), 'Models of Construction Innovation', *Journal of Construction Engineering and Management*, vol. 124, no. 3, pp. 226-231.

Teece, D.J. (1989), 'Inter-organizational Requirements of the Innovation Process', *Managerial and Decision Economics*, special issue, John Wiley and Sons, New York.

Winch, G.M. (1998), 'Zephyrs of Creative Destruction: Understanding the Management of Innovation in Construction,' *Building Research and Information*, vol. 26, no. 4, pp. 268-279.

Chapter 9

Skills and Occupational Cultures

Rob Shields

This chapter explores two aspects of the dissemination and successful take-up of innovations – work cultures and skills. The focus of much of the innovation literature is on product development strategies and processes, such as partnering with other firms to develop market and consolidate specific innovations. However users' skill level, aptitude for new skills, and previous experiences of adopting innovations are factors in the successful take-up and dissemination of major innovations. In the construction trades, we find strong work cultures anchored around traditions of specific tasks and technologies (the trowel, mortar and brick, for example) independent of direct control. These occupational cultures might be expected to be more significant issues in construction innovation than in other industries such as manufacturing where workers' tasks are assigned directly by managers.

This chapter first considers the emergence of views that 'culture matters' and calls for changes in the 'culture of construction'. Understandings of culture used by management studies and analyses of construction will be critiqued in favour of the sociological understanding of culture as repeated routines and practical habits, developed in the work of Pierre Bourdieu and others. Understanding culture as practices as well as values makes its relevance to innovation much clearer. Skills and competencies are clearly related to routine practices of organizations and occupations. The conception history and on-site understanding of skill in the construction trades is considered in relation to challenges of innovation and the needs for training related to new technologies. The politics of skilled labour and innovation will be considered before turning to problems of skill shortages, quality and productivity. Problems of inadequate statistical measures are discussed as well as implications of the apprenticeship system for the take-up of new and safer tools and technologies.

9.1 Work Culture and Innovation

Proponents of the view that 'culture matters' in innovation argue that it is,

> necessary to differentiate between a technological and a social 'face' of innovation. The first considers product and process innovations as its focus of analysis. The latter tries to explore the dynamics of organizational change at

the firm level and various aspects of human resource management at the functional level (Hauser 1998:240).

Organizational culture is often mentioned as a key determinant of the innovativeness of firms because it sets the stage for other social aspects of innovation processes. In recent studies of barriers to innovation and competitiveness culture is frequently cited. In the case of the UK construction industry, Latham encouraged the construction industry to foster improved relationships with clients and their industry partners including subcontractors (Latham 1994). In Egan's *Rethinking Construction,* changes in working conditions, skills and training require 'Substantial changes in the culture and structure of UK construction' (Egan 1998 online: Ch. 4 ss.51). To move away from a confrontational culture, 'cultural and process improvements' (Ch. 4 ss.62) would involve 'a culture of radical and sustained improvement' (Ch. 4 ss.72), 'of trust and respect' (Ch. 6 ss.84).

In the construction innovation literature key works relevant to organizational culture include the results of a CIB task group on the topic edited by Fellows and Seymour (2002). In this report, Barthorpe provides an historical overview of the term (Barthorpe 2002). The emergence of Human Resource Management as a response to the quantitative approach of Taylorist 'Scientific Management' came together with competitive pressures on firms in the 1980s to create a heightened interest in corporate cultures and their role in national competitiveness. Burack defines corporate culture as 'The way things are done in organizations'. He emphasizes 'shared assumptions, beliefs and values...behavioural norms and expectations' as 'the glue that holds the corporate community together (Burack 1991 cited in Barthorpe 2002:12; for a survey see Brown 1995). As opposed to national culture, Schein defines organizational culture as a pattern of assumptions that is taught to new members as the correct way to perceive, think and feel in relation to problems that the organization faces (Schein 1992 cited in Barthorpe 2002. Other studies of firms show that culture has become a by-word for values and the atmosphere which hinder or enhance corporate performance and staff retention in companies (e.g.Wilkinson and Scofield 2002: 72; Tayeb 1991).

Hofstede's international study of IBM in the 1980s identified four organizational cultural differences which varied across nations: power distance; uncertainty avoidance; individualism-collectivism; masculinity-femininity (Hofstede 1991). Illustrating the difficulty that encyclopedic definitions have, a fifth dimension, 'Confucian dymanism' or long-term/short-term orientation, was added after a consideration of Chinese values (Bond 1987 cited in Barthorpe 2002:18). Other large studies also focus on values and attitudes (Deal and Kennedy 1982; Trompenaars and Hampden-Turner 1998). More recent studies analyze the convergence or divergence of values amongst managers in multicultural firms (e.g. Selmer and de Leon 1996). Hauser provides a synthesis of the English and German literature (1983) that finds the same problems with this approach. One study concludes,

We think people do have implicit images based on their culture of what an organization is and we think these images differ fundamentally, although it is not easy to describe them accurately. We also think people have implicit images about what their markets are. These images disclose their assumptions about how to operate effectively in these markets (Neuijen 2001:28).

This comment is telling. 'Images' of organizations and markets are based on what their perceived role and capacities are. The problem with all these studies is that they try to generalize from values and norms to the way things are done. That is, they derive action from belief rather than studying action directly. This is seductive but is a conceptual shortcut which masks assumptions, unexamined stereotypes and short-circuits research. For example, one might assumes that individual roles and behaviours are aligned in a one-to-one manner with overarching values of construction companies such as entrepreneurship, however the situation is more complex. Workers on site do not necessarily share in corporate values. These are always tempered by competing, and sometimes countervailing, values geared around union and journeyman traditions, the increasing individualization and atomization of the workforce and personal identity (masculinity, ethnicity or subculture, and so on).

New Zealanders have long been associated with a "she'll be right" attitude (Hofstede 1980). What this means is that people are known to take unnecessary risks. Unfortunately, this culture has proliferated into the area of safety on the construction sites and has led to a high accident rate... (Wilkinson and Scofield 2002:69).

An unnecessary assumption is made that culture is a property of nation-states rather than an attribute that belongs primarily to ethnic groups or varies across national populations. Thus one might contrast class cultures, the cultures of particular regions or cities, or the cultures of occupational groups that coexist within one firm or on one site. All these are typical examples found in the social science literature.

9.2 Culture as Practices

Outside of the management and human relations literature, culture is less often treated as only a set of beliefs and values. It is understood by the social sciences as the 'whole way of life' including practices and routines. Although the precise definitions and terminology varies, this anthropological understanding is a common origin for many definitions (Kroeber and Kluckohn 1963:81). It is applied to subsections of societies in the analysis of occupational and local communities and regional identities. In practice, culture is therefore an organizing logic that relates practices to beliefs. This middle level has generally been missing in discussions of culture in relation to construction and engineering. The 'stuff' of culture lies in 'how' even more than 'what' is done on an everyday basis. Everyday procedures include both the routines of working together and the boundaries of cooperation - groups that are not worked with or are seen as outsiders and others For this reason culture is often characterized as involving something of a 'style', a traditional way of going about a task, or a habit. When researchers investigate work cultures in firms or in a trade, they often seek the logic that links stories of 'how we do things around here' together with each other and with values or formative historical events (Becker 1974).

Stories of 'how we do things around here' are both the basis of solidarity amongst workers as well as explanations of the organization of work – and they are relevant to an innovation which must find its place within an evolving organization and style of working together effectively (Shove, Packwood, and Shields 1997). This constitutes a framework for problem solving. Pierre Bourdieu refers to this set of predispositions which guide improvizations in daily routines as the 'habitus' - a durable set of predispositions related to a clearly defined 'social field' (such as building activities) and forms of capital, including economic cultural and social capital (Bourdieu and Passeron 1977:76; Bourdieu 1977:81-2).[1] Bourdieu often refers to habitus as both a set of algorithms or rules of thumb as well as something more like an attitude and outlook (Bourdieu 1984:424). Habitus is the *propensity to choose one approach or one way of working over another*. This is the level of the organization of work, which changes when innovations are introduced. Habitus provides a social and cultural explanation of the observation that most respondents to surveys of culture comment idiomatically not on values but on activities done in a firm ('we all go to the annual trade fair', or 'its always been done that way') in response to questions about firms' cultures.

Bourdieu's understanding of culture as habitus is by now one of the best-known sociological theories. It is frequently drawn upon in work on learning and knowledge in organizations and in studies of communities of practice (Wenger 1999; Delamont and Atkinson 2001; for a survey and critique see Mutch 2003). His specific contribution is to treat habitus as stratified across societies rather than simply changing over time. It is 'deep' in that it extends from the micro level of gestures (how a tool is used 'correctly' and safely) to the macro level of collective ideologies (values of independence, for example (Bourdieu 1977:91)).   It is different from most organizational and business descriptions of culture, but typical of many sociological and anthropological approaches developed over the last 40 years.[2]

Although students of organizational culture argue that, 'in non-routine processes like innovation processes human beings are faced with missing orientation patterns' (Hauser 1998:239), Bourdieu's theory of habitus provides an explanatory framework for the selection of problem-solving alternatives when workers are confronted with a new situation or innovation. It also explains the tendency to fall back on 'tried and true' methods even in cases of pure trial and error.

A second contribution is to direct attention to culture as repeated routines. Habitus is practical knowledge (Bourdieu 1984: 466). It continually 'adjusted to the particular conditions in which [it is] constituted' (Bourdieu 1977:95). This approach suggests that certain organizational habits are cultivated and durable even though they must be continually flexible. These might be actions, such as the way of communicating problems encountered in assembly processes. Or they might be values such as attention to detail and a stress on quality. A strength of understanding culture as habitus is that routines can be observed and documented, whereas values must be inferred, making them less amenable to research. By contrast, studies of the role of organizational culture in innovation often begin from the premise that,

> Traditional planning, organizing and standard procedures are marginal aspects if the research goal is the explanation of those factors influencing innovation processes. Innovation *per se*...can be planned only to a limited extent (i.e. development) and management is mainly restricted to create innovative contexts...as a.... situation culture, as a substitute for structure and as a complement to management and leadership... (Hauser 1998:240).

Because persistent elements such as tradition and procedure are integral to habitus and to a deep understanding of organizational culture, management studies approaches may eliminate the elements of culture relevant to innovation researchers. Hauser himself goes on to acknowledge this problem when he critiques approaches that reduce organizational culture to values such that, 'Culture is defined as a set of shared values, norms, and knowledge within a firm' (Hauser 1993:240 see also, for example, Deal and Kennedy 1982; Kotter and Heskett 1992). But there is not much else left once one eliminates all tradition and routine procedure. These are not values but patterns, ways of doing things that constitute part of the common knowledge of individuals in an organization or trade. This shared reference point is flexible but constitutes a context for problem-recognition and solving, and for estimations of value. It is an extension of practical knowledge – a context that prompts innovation, and coherent set of practices, ideas and beliefs innovators cannot ignore if they expect new products or processes to be integrated into a social field such as an occupation or trade. Without any element of action, which also includes creative activity, it is not surprising that researchers 'are not able to explain the function of culture within the innovation process in general' (Hauser 1998:240).

Bourdieu makes a third contribution by tying habitus to domains of action or 'social fields' and to forms of capital (social, cultural, economic). Habitus is a set of social dispositions to spontaneous and creative action based on one's position in a 'field' (a group of inter-related social actors) and possession of various types of capital (such as resources, cultural power, social networks). It is on this basis that the construction site and engineering projects might be seen as the sites of intersections of occupational, class, ethnic and regional habitii.

Gunderson notes that construction is different from other industries in that a large number of intermediaries separate workers from prime contractors. As a result, labour tends to be seen by those managing projects as only a cost, which is to be minimized (2001:7). A second feature of this structure is that even direct employees identify with their trade or craft, rather than with their employer, setting the stage for communication difficulties between the two when innovation is attempted. This is a key feature of the habitus of construction in general and each trade can be profiled along a continuum of relative autonomy from the project manager's direct control. Third, as noted in other chapters, construction is sensitive to economic cycles because clients can often defer new construction. Thus there is a high economic risk and rates of bankruptcy. The turnover in firms disrupts long-term employer-employee relationships and leads to a plethora of informal, personal bonds based on individual reputation and past favours. Fourth, widely distributed, temporary construction sites are difficult to police. As a result there substantial rates of seasonal, informal and 'underground', unskilled and inexperienced employment in the industry have been consistently found (O'Grady and Lampert 1998; Gunderson 2001:8).

Although Bourdieu does not focus on innovation, he discusses the take-up of technologies such as photography (Bourdieu et al. 1990:14). The use of tools is indicative of the relevance of habitus to an understanding of the use of technologies. Rather than opposing technology to society, tools and technologies are integrated with habitus. Ailments such as repetitive strain injuries illustrate that tools demand that fixed actions or gestures be carried out as a precondition of producing the expected or orthodox results efficiently. These conventions also limit what can be done with tools or technologies, imposing constraints on, for example, the temperature at which concrete pouring can be done out of doors. More broadly, this approach understands innovations not as 'things' that fulfill needs but objects and procedures that are socially shaped along with their meanings and their roles or the uses to which they are put. Similarly, the analytical categories in which we group technologies are derived from the usage to which they are put – practices which follow their own logic of the habitus, not the abstract logic of engineering or product design (Sterne 2003: 374).

Rather than formulating innovation as a break in habitus it may prove more useful to conceive of innovation as involving not only the reorganization of products and work processes but as a process which requires that people reorganize their own propensities to improvise in certain ways rather than others when problems are encountered. The tendency, however, is to carry an older habitus over to operating the new technology, tool or process – hence web 'pages' and file 'folders' hark back to an earlier era when information was organized and presented in print media.

Carrero *et al.* argue that innovation on work sites arises through learning and adaptation (2000:489). Their focus is on the onsite process of implementing innovation rather than a decision to innovate that may arise from management strategy or sources external to the construction site. Innovation is viewed as having '...a lineal development, with easily identifiable and predictable sequences' (Carrero, Peiro, and Salanova 2000: 491). These dynamic 'process' approaches offer an advance on research that ignored the importance of innovations becoming successfully embedded in flows of work processes. However the impediment of off-site training outside of the control of firms, suppliers or clients is less discussed (see Shields and West 2003).

It may be that negative attitudes to change are derived from the way in which innovation is implemented on construction and engineering sites. Many innovations are not primarily related to buildings or to built infrastructure, but are ergonomic innovations to improve health and increase safety. Health and safety innovations are often only adopted as a result of regulatory intervention. While activists and unions may lobby for such changes, they require a reworking of the day-to-day routines and physical comportment onsite. The rapid adoption of ergonomic and health and safety innovations has not be aided by the lack of awareness of the importance of habitus and the extent to which even obvious innovations such as safety harnesses my discomfort, impede and slow down individual workers who consequently resent the very innovations which are most in their own interest.

9.3 Skills and Culture

The Oxford English Dictionary defines 'skill' as firstly 'the power of discrimination...a sense of what is right or fitting....' and secondly the 'capability of accomplishing something with precision and certainty; practical knowledge in combination with ability.' Skills are often thought of as things, and managed in the form of qualifications and certificates. But both the semiotic and the ergonomic researcher approach skills as capabilities that are the outcomes of more tangible processes such as specific tasks, cognitive operations or physical actions. Like culture, skill is an overarching term whose meaning can only be understood through analysis of its components. Over the course of the 20th century beginning with the research of Taylor, it has become routine to analyze factory manufacturing and assembly, and warehousing and shipping operations with the objective of breaking down traditional notions of skill into sequences of gestures or actions. The intangible qualities of a notion such as skill are thus eliminated in favour of more tangible and thus more manageable actions.[3] The early twentieth century example, Taylorism, aimed to improve the efficiency of assembly line work in particular by identifying which aspects of tasks could be done at each stage of assembly so that complex assembly tasks could be parceled out to separate, unskilled workers each responsible for a repeated set of maneuvers. Getting rid of the need for a 'sense of what is right or fitting' does not deal with questions of quality, which only becomes a more acute issue for managers.

The skilled trades, by contrast, have maintained a fidelity to their craft history. A perception of construction as the 'industry which time forgot' has been based on the nature of the work itself (Sobel, 1995, 48). Labour intensive, involving manual skills, physical strength and with an outward appearance of little organization or oeven chaos on the work site itself, construction trades have only been the focus of research in the last decade.

> Given the inherent 'craft' basis of most construction work...[it is]...the general view that building trades have been insulated from the impact of technological change and work re-organization that the manufacturing and service sectors have witnessed (Sobel 1995, 48).

Skills are an integral part of the habitus of the construction trades and bear directly on questions of productivity. They are developed and taught not as 'skills' abstractly divorced from real-life situations but are the elementary particles of work situations and habitual routines. They lie at the heart of efficient and effective work processes. For example, one does not learn how to use an air hammer but rather 'lays flooring' and develops the relevant precautions, actions, muscular ability, expectations of work-mates and institutions (such as suppliers, or other trades) and an simultaneous understanding of both what constitutes high quality and how that specific level of tolerance of error is achieved. This also includes shortcuts and more effective ways of working unimagined by suppliers and innovators. It also involves acumen developed through practice at, for example, coordinating hand and eye. At traditional example would be the pressure versus the speed at which a trowel should be moved to achieve a good plaster finish. All of these examples fit directly into Bourdieu's notion of habitus, defined in relation to a given social field (e.g. a construction trade) and cultural capital (certification, seniority).

On-site, 'skill' is understood in terms of the ability to accomplish certain jobs, all of which involve several tasks within specific environments or contexts. These clusters of tasks and aptitudes are best described as 'competencies' – a middle level term between the abstraction of 'skill' and the concreteness of 'task'. For example a pipe fitter may practice for and be tested on welding together two types or sizes of pipe but this requires both preparation of the materials and equipment, welding and clean up. The welding itself is only a small part of the overall procedure parts of which are now sometimes done by an automated welding machine. With such innovations, 'perfect' welds are possible, but the pipe fitter/operator still needs to recognize whether or not the machine has functioned well and created a good weld. New computer skills are required and in some cases non-destructive testing techniques must be learned (ultrasound, infrared and computerized microscopic examination techniques, see Shields and West 2003).

The welder's skills – plural – in accomplishing this range of activities are examined by setting standardized tasks for apprentices. One attempt to change this system is to break up the system of apprentice and journeyman by modularizing trades certification. For example, rather than waiting until the completion of a long apprenticeship, an electrician could be licensed to perform domestic wiring work while still in the process of learning how to work on more complex electrical systems. With changing technologies this has already started to happen in a de facto manner as trades people have had to take specialized courses to work with new materials (fibre optics, plastics and ceramics) and new tools (motorized equipment requiring licenses and robotic tools). The rationale for 'modularization' is the long time it takes to train apprentices (often five years or more) in the face of shortages caused by an aging population. The average age of many skilled trades in Canada is over 50 at the time of writing.

However, another aspect of skill is not directly addressed by modularization projects. This is the understanding of quality and a rationale for particular approaches (insisting for example that an electrical outlet must be placed at a certain level from the floor, even if it is inconvenient to users). Building and electrical codes may forbid or prescribe the location of some outlets but the strong sense of the 'rightness' of such practices can only be established by immersion in the culture of electricians. This culture is communicated in tale tales and moral stories of accidents and dangers. Again and again on sites, skill is summed up in pranks played on apprentices, persuasive anecdotes told on breaks, and hair-raising tales of narrow escape that make crystal clear that no other alternative is to be considered. The only surprise is that innovators generally fail to address trades via this medium of authoritative folk tales. Instead, an appeal to the masculine stereotype of most trades only reinforces skills shortages by reproducing the gendered habitus of the construction and civil engineering.

Technologies tend to concretize and sediment social organization and power relations. Different groups use different technologies in different ways at different moments. However these social characteristics (such as the gendered use of technologies) may change over time (see Jensen 1988). In contrast to rather positive perspectives toward innovation that focus on the firm, the impact of technological change and manufacturers' innovations on labour in the construction industry has often generated a negative response from individuals and unions. In particular, process innovations (defined in earlier chapters as changes in method or the procedure of product application) often aim to reduce the demand for skilled labour by two means. First, through the reduction in time and the number of labourers required to complete a task. Second, through the marginalization of skilled labourers needed to accomplish tasks (Slaughter 1993; Sobel 1995). The diversity of interests around innovation in general may be recognized by the industry, but has not been fully acknowledged in construction research or in industry-government fora. One rarely finds representatives of construction labour present at symposia on innovation. Process innovation has often been implemented to the detriment of workers. Speaking of Central Canada, Sobel notes that,

> ...a whole new array of building technologies, materials, systems, and products are also responsible for the decline of employment in the building trades. The few projects currently underway in southern Ontario require fewer workers than before, and they are brought to completion more quickly than ever. The same or more construction work can be done with fewer workers because of technological change and industry restructuring (Sobel 1995, 49).

In some cases, skills have remained relatively unchanged but new mechanical devices, such as lifts and jacks allow individuals to do tasks that once required teams to do. For example, the assembly and lifting into position of ductwork along a ceiling can be done by a single person aided by small, remote controlled tractors (known under their trade name as 'Genie lifts') that lift the ducts to the ceiling on extendable, hydraulic booms. However, the result is a loss of human team contact, increased individual responsibility, and possibly stress, loneliness and loss of an appreciation of one's contribution as a participant in a larger process.

9.4 Power, Innovation and Skilled Labour

Innovation goes hand in hand with both maintaining and restructuring power relations between actors in a given context. In construction and engineering, these may be the relationships of small and large firms, between project managers and contractors, those within the industry, and those between labour and management within firms. Both product and process innovations have been implemented top-down in an attempt to 'modernize' what is often perceived as an industry lacking in forward-looking technology and practices, and thus is seen as lagging behind other sectors of the economy. Faced with a drop in construction activity in the early to mid 1990s, new forms of work organization, increased mechanization and the resultant decrease in the need for 'man power' have attempted to increase the productivity and efficiency of construction to enable its competition and viability in a changing economy (Sobel 1995). The result for the labour force has been a process of labour marginalization, 'deskilling', disempowerment, and labour displacement through changes in the industry, as workers compete for jobs, and firms compete for a declining number of construction projects.

Braverman famously identified the process of deskilling which accompanied the application of Taylorist principles and the use of automation in workplaces. Individual workers felt they became servants to machines, while the required level of certification as well as ability was reduced. The prestige of trades was diminished by their loss of responsibility for and control over the work process. For others, a process of re-skilling took place as tasks were reorganized into new job descriptions for para-professionals (including designers) and technicians (such as 'installers' and a variety of other technical service positions) who took on non-traditional tasks and learned the associated skills and abilities (Braverman 1974).

Historically there are many examples of technologies replacing rather than complementing skills (see Mokyr 1990). However, most new technologies of the late-twentieth century appear to be skill complementary by design boosting the 'college premium' for skilled workers (Acemoglu 1998:1056). An increasing net level of skills and training of a workforce is closely tailed by increases in expectation and demand for these increased skills. Comparing across both high technology (computing) and construction and engineering in 7 OECD countries, as the pool of skills increased, demand for skilled labour similarly increased.

> Most technologies once invented are largely nonrival goods. They can be used by many firms and workers at low marginal cost. When there are more skilled workers, the market for skill-complementary technologies is larger. The

inventor is therefore able to obtain higher profits, and more effort will be devoted to the invention of skill-complementary technologies (Acemoglu 1998:1056)

Using measures of innovation and technical change based on surveys of industry research and development expenditures, others also conclude that 'skill-biased technical change is an international phenomenon that has had a clear effect of increasing the relative demand for skilled workers' (Machin and Van Reenen 1998:1217). This situation may also prevail in construction, but low R&D expenditures in the industry require different approaches be taken.

One important impact in OECD countries has been a lack of interest amongst youth in qualifying for the skilled trades, preferring instead positions associated with the higher status command and control roles in the project process. The shift in status and power (understood in the contemporary sociological sense not simply as force but as influence or the ability to act upon others' actions) has affected the capacity to innovate at the level of on-site activities and functions. On-site activities are reduced to 'processing' and to installing. With the reduction in control over their jobs, the tasks and processes by which jobs were accomplished, an important set of aptitudes was lost from the habitus of each of the skilled trades. What was once a daily necessity was lost. Suppliers and designers attempted to save time by eliminating the need for on-site worker problem-solving with a view to the quality and long-term durability of the final product. Henceforth what didn't work or didn't fit or was an obvious error became the responsibility of the professionals and paraprofessionals – if workers bothered to alert them to the problem. As part of this process, input from tradespeople or other onsite actors may be too easily denigrated or dismissed by engineering and construction site managers. New products demand new processes, making innovation a challenge because *not only problem-recognition and problem-solving but team-work skills have been lost* from the 'culture' of the construction site as whole.

9.5 Innovation and Skills Shortages

Skill shortages were predicted as the prestige of trades declined, but the changing demographics of those working in the industry in OECD countries have exacerbated this problem. The average age for the construction workforce in Ontario Canada is over 50. In Europe high rates of turn-over in the construction workforce threaten levels of skill as workers retire, change to less physically demanding types of work or move within and outside of Europe (Lebeau and Vinals 2003: 9). There may be technical skill shortages but debates on skill and labour shortages may be indicators not of technical problems but of shortages of workers who combine both technical skills and what have come to be known as 'essential skills' such as interpersonal, and teamwork skills. In even more demand are those who combine these two areas with managerial skills such as numeracy and ability to work with budgets. But many analysts have argued that there is little evidence of widespread skill shortages. Rather, in Central Canada, shortages are found only in very specific regions and specific trades such as residential carpentry and drywalling (Ontario Construction Secretariat, Peartree Solutions, and Prism Economics and Analysis 2003). There is also a political dimension to this debate. The rhetoric of shortages of skills and labour favours industrialization (such as prefabrication), changes in labour relations around mechanization and flexibilization (as workers become employees rather than self-employed tradespeople) and the opening of borders to workers drawn from overseas (drywallers from Chile working temporarily in Canada, for example (see Bertin 2002)). However, these may or may not be appropriate solutions. As will be discussed below, there is a general lack of information that would inform policy options.

The perception that the skill level of the construction workforce cannot be counted upon is significant for innovation. Firms avoid technologies and approaches that workers are not experienced with. Specifiers and contractors look for an established track record of success with particular product or system amongst subcontractors and trades.

It is difficult to implement technical change requiring retraining in situations where control over the execution of work has been entirely delegated to subcontracted labour, as in the case of certified trades. Studies indicate that training strategy compliments and may even be the innovation strategy of firms (Baldwin 1999). But in surveys of construction firms, education and training are generally not considered a significant productivity factor (re. Canada see Seaden et al. 2001:32). However, Seaden et al suggest that positive attitudes toward training were correlated with – and are possibly a precondition for – innovation by firms (2001:62). Albers et al note:

> Skilled workers in the building and construction trades are recognized as having a distinct occupational culture characterized by the positive control they exert over their job tasks. This work autonomy is related to the skills learned while serving their apprenticeship, their tool ownership, and the "portability" of their skills (Albers et al. 1997: 644).

Control over quality and skill is delegated to workers. When unions are responsible for training, American research has suggested that there may be a 20-30 per cent boost to productivity but unionized wages tend to be higher by the same amount (Allen 1986 cited in Lebeau and Vinals 2003:194). Besides the obvious reasons such as training and standardization of skills held by certified tradespeople, the increased cost may induce general contractors to pause to consider and more thoroughly plan the project to be undertaken.

On the other side of the coin, there are increased costs when certified tradespeople are required by unionization rules to perform very simple tasks that could easily be accomplished by an unskilled worker. Also, especially where workers are self-employed subcontractors, increases in productivity require the intensification of work within established, certified technologies as the basis of competitiveness. Technological innovation that requires different work processes and new operational skills becomes a challenge (Winch 1998). Control over skills and worker autonomy has implications for innovation in construction because there are more independent gatekeepers who are able to refuse change unless it is in their interest.

As a result, retraining has not tended to be seen as a major area of interest for the construction trades. Rather than an ongoing process, training has been associated with apprenticeship, a status to be left behind. Training also represents time not working and therefore unpaid time for most certified trades working on diverse projects rather than for a single employer. However, Barnett and Storey (2001) argue that training is intrinsic to innovation and sustainable productivity growth within an industry. They treat innovation as the process of 'innovating' rather than a static conceptualization of innovation (2001: 84). The development of workers and the creation of a local skill base of future workers are integral to the competitiveness of firms in that locality. For the case of skilled construction trades, however, training is coordinated by unions with a view to standardizing the offerings and output of their members who are allocated to worksites on a day-to-day or job-to-job basis under the management of the union's business agent. These unions own and operate training facilities and accept or certify course taken at state colleges and polytechnics. Innovation may occur on site through the diversification of product applications, and the cultivation of a work culture open to change, questions and challenging of both products and process in order to create new and more effective ways of doing things (Barnett and Storey 2001: 92-93).

The lack of information on construction and engineering labour is a contributing factor to the persistence of skill shortages. Most of the existing data and models are severely limited when it comes to predicting the demand for and potential shortages of skills, especially in construction. Some national governments have discontinued national surveys of employers in the industry or do not disaggregate employment data along industry lines (such as Help Wanted Index and similar surveys). Some analysts argue that occupational projects supplemented by qualitative assessments are more useful (Gunderson 2001) despite their reliance on employer surveys and investment only in new construction rather than renovation, installation and refitting and they do not incorporate factors such as immigration, mobility or retirement. The number of small employers and informal practices poses a challenge to any assessment of the workforce, especially in residential construction.

One approach has been the creation of industry advisory councils, such as the Canadian Construction Sector Council which in the absence of macroeconomic data has aimed to bring together stakeholders and well-informed players 'close to the ground' in different areas of construction. The objective has been to provide an information clearing house for matching demand for qualified tradespeople with supply of skilled workers; to encourage flexibility and responsiveness in apprenticeship systems; to increase the currency of skills and to facilitate worker mobility in response to regional demand (Construction Sector Council 2003).

Shortages of skilled workers and construction labour in general has important economic implications because of the linkages between all forms of construction (renovation, repair, new construction) and other activities, not to mention problems where shortages of one skill impact on other, complementary or dependent trades from working. Social implications include general quality problems and hazards to health and safety. One response is either further bifurcation of construction into prefabrication of new construction and an informal sector of repair and renovation and a DIY sector (do-it-yourself) providing consumer kits and packages for specific jobs (Gunderson 2001:90). This is significant in that the renovation and DIY sector is the largest value of residential construction in some countries. Prefabrication is of growing importance as construction becomes a global industry. Lack of skilled and experienced workers in major markets such as China have led designers and engineers to favour manufactured building components, turn-key units which can be easily installed and products which feature single-coat applications (see Lebeau and Vinals 2003:216).

The changing demographics of the construction labour force have contributed to a new stress on health and the security of workers – a so-called safety culture. But the arrival of younger workers with different expectations, a familiarity with push-button, machine-centred work and a willingness to take risks but less tolerance for harsh conditions has resulted in a demand for ergonomic and health and safety innovations in construction. These are both product innovations (for example, new, ergonomic, safety vests) and process innovations (procedures which isolate workers from dangerous levels of pollution (noise and chemical) and remove some from the project floor entirely (courtesy of mechanization in general and what has been referred to as a 'culture of cranes' (Shapira 1996).

However, the goals and methods of apprenticeship are based on socialization into an occupational culture, copying and adopting the approach of more senior mentors. Albers et al identify this as a barrier to innovation by new generations of entrants into trades who may be more aware of ergonomic risks or new tools (curved-handled hammers, knee-pads, safety harnesses, for example). Fear of ridicule is paralleled by other barriers to innovation, such as employers' unwillingness to spend funds on devices such as simple lifts and jacks for lifting drywall sheets or other finishes into place (rather than relying on sheer muscle power). In the absence of communication and training approaches which would reach experienced workers, many innovators and observers of the industry have relied on government regulation to force innovation on construction and civil engineering (Albers et al. 1997:645).

9.6 Conclusion

Construction is well-known for the strong identification of workers with their trade rather than with firms. Calls for change in the 'culture of construction' are undermined if 'culture' is relegated to questions of values and beliefs. In this chapter, a more operational approach was introduced which focused on routines and practices. Habitus, sets of practical routines and dispositions toward certain ways of solving problems encountered in construction projects, was suggested as a useful approach to the cultures of construction trades. This was proposed to solve the problem of trying to generalize from values and beliefs to productive activities. Also habitus provides a framework for understanding the propensity or resistance to change amongst specific occupational groups.

Habitus raises further questions beyond the scope of this chapter such as whether or not it is possible to train construction personnel for innovation. Problem-recognition and problem-solving across novel and proven components and systems and team-working skills which might allow trades to cooperate with each other and with managers are needed. Could construction and engineering projects become more collaborative endeavors? In other industries, research has shown that a work culture open to change, and workers who question and challenge both products and processes in order to create new and more effective ways of doing things are advantages to firms (Barnett and Storey 2001: 93). Certain skills seem to be crucial to the practical problems of any innovation – can they be reintroduced to the construction culture? Can they be taught?

Skills and competencies tie questions of habitus to productivity and to the challenges which come with change and innovation. But innovation goes hand in hand with the perpetuation and restructuring of power-relations between the trades and between skilled labour and management, as well as between small and large firms, and between project managers and contractors. The relationship between innovation and skilled labour has been shown to be more complex than the stereotyped view that innovation always threatens the interests of workers. Successful innovations either attend to the reality of the habitus of the various actors and the culture of construction that they create as a whole or confer enough value-added for clients and competitive advantages for suppliers that they become factors that reshape the industry as a whole. Some technologies which are skill complementary by design, and technology-related retraining offer opportunities to enhance the capacity of firms to innovate further. However, the effectiveness of providing these skills via training only under apprenticeship systems or only by unions alone rather has been questioned.

Notes

1. Thus value is not merely economic but exists in several registers. Cultural capital, such as expertise of a qualification, can be converted to economic capital. Furthermore, economic capital is partly constituted by cultural agreements, norms and practices. Thus these forms of capital are not counterposed to each other but mutually constitutive of value. This approach is developed further in the work of Boltanski and Thevenot (see Boltanski and Thevenot 1991; Boltanski and Chiapello 1999).
2. Habitus can also be found in Marcel Mauss' inter-war studies of the practical knowledge which forms our embodied subjectivity (knowing how to dig with a spade or throw a ball, for example (Mauss 1979)) and Norbert Elias' 1939 discussion of historical

changes in manners and techniques associated with social norms of civilized behaviour (for example, the fork as a technique of eating (Elias 2000)).
3. On the importance of differentiating between tangible things like a given task and intangibles like 'skill', see Shields 2003.

References

Acemoglu, D. (1998), 'Why do New Technologies Complement Skills? Directed Technical Change and Wage Inequality', *Quarterly Journal of Economics,* vol. 113, no.4, pp. 1055-1089.
Albers, J., Yuhua, L., Lemasters, G., Sprague, S., Stinson, R., and Bhattacharya, A. (1997), 'An Ergonomic Education and Evaluation Program for Apprentice Carpenters', *American Journal of Industrial Medicine,* vol. 31, pp. 641-646.
Allen, S. (1986), 'Unionization and Productivity in Office Building and School Construction', *Industrial and Labour Relations Review.* vol. 39, no. 2, pp.187-202.
Baldwin, J. (1999), 'Innovation Training and Success', Statistics Canada, Micro-Economic Analysis Division, Statistics Canada, Ottawa.
Barnett, E., and Storey, J. (2001), 'Narratives of Learning, Development and Innovation: Evidence from a Manufacturing SME', *Enterprise and Innovation Management Studies,* vol. 2, no. 2, pp. 83-101.
Barthorpe, S. (2002), 'The Origins and Organizational Perspectives of Culture', In. *Perspectives on Culture in Construction,* D. Seymour and R. Fellows (eds.), Paris, CIB.
Becker, H.S. (1974), 'Art as Collective Action', *American Sociological Review,* vol. 39, no. 6, pp. 767-776.
Bertin, O. (2002), 'Builders Suffer Labour Shortage: Temporary Visas Help Ontario Fill Skilled Worker Void', *Globe and Mail,* March 5, Online Archive.
Boltanski, L. and Chiapello, E. (1999) *Le Nouvel esprit du capitalisme.* Gallimard, Paris.
Boltanski, L., and Thevenot, L. (2000). 'The Sociology of Critical Capacity', *Zhurnal Sotsiologii i Sotsialnoi Antropologii: The Journal of Sociology and Social Anthropology*, 3(3).
Bourdieu, P. (1977), *Outline of a Theory of Practice,* Translated by R. Nice, University Press, Cambridge.
_____. (1984), *Distinction: A Social Critique of the Judgement of Taste,* Routledge, Kegan Paul, London.
Bourdieu, P., Boltanski, L., Castel, R., Camboredon, J-C., and Schnanner, D. (1990), *Photography: A Middle Brow Art,* Translated by S. Whiteside, Standford, University Press.
Bourdieu, P., and Passeron, J-C. (1977), *Reproduction in Education, Society and Culture,* Sage, London.
Braverman, H. (1974), *Labour and Monopoly Capital: The Degradation of Work in the Twentieth Century,* Monthly Review Press, New York.
Brown, A. (1995), *Organizational Culture,* Pitman, London.
Burack, E.H. (1991), 'Changing Company Culture: The Role of Human Resource Development', *Long Range Planning,* vol. 24, no. 1, pp. 89-95.
Carrero, V., Peiro, J.M., and Salanova, M. (2000), 'Studying Radical Organizational Innovation through Grounded Theory', *European Journal of Work and Organizational Psychology,* vol. 9, no. 4, pp. 489-514.
Construction Sector Council. (2003), *Construction Industry Human Resource Challenges and Responses: Meeting Human Resource Needs,* Construction Sector Council/Conseil

Sectoriel de la Construction, [Cited June 2003]. Available from: http://www.csc-ca.org/pdf/HR_rep_e.pdf.

Deal, T.E. and Kennedy, A.A. (1982), *Corporate Culture: The Rites and Rituals of Corporate Life,* Cambridge, Mass, Addison-Wesley.

Delamont, S. and Atkinson, P. (2001), 'Doctoring Uncertainty: Mastering Craft Knowledge', *Social Studies of Science,* vol. 31, no. 1, pp. 87-107.

Egan, J. (1998), *Rethinking Construction,* Department of the Environment, Transport and the Regions, London.

Elias, N. (2000), *The Civilizing Process,* Translated by E. Jephcott, Blackwell, Cambridge.

Gunderson, M. (2001), 'Economics of Personnel and Human Resource Management', *Human Resource Management Review,* vol. 11, no. 4, pp. 431-452.

Hauser, M. (1998), 'Organizational Culture and Innovativeness of Firms: An Integrative View', *International Journal of Technology Management,* vol. 16, no. 1-3, pp. 239-255.

Hofstede, G. (1980), *Culture's Consequences: International Differences in Work-Related Values,* Sage, London.

_____. (1991), *Cultures and Organizations: Software of the Mind,* McGraw-Hill, London.

Jensen, J. (1988), 'Using the Typewriter: Secretaries, Reporters, and Authors 1880-1930', *Technology in Society,* vol. 10, pp. 255-266.

Kroeber, A. and Kluckohn, C. (1963), *Culture: A Critical Review of Concepts and Definitions,* Penguin, New York.

Latham, M. (1994), *Constructing the Team,* London, Department of the Environment.

Lebeau, D. and Vinals, J. (2003), *Bâtir et innover: tendances et défis dans le secteur du bâtiment,* Conseil de la Science et de la Technologie, Quebec City.

Machin, S. and Van Reenan, J. (1998), 'Technology and Changes in Skill Structure: Evidence from Seven OECD Countries', *Quarterly Journal of Economic,* pp. 1215-1244.

Mauss, M. (1979), *Sociology and Psychology: Essays,* Translated by B. Brewster, Routledge and Kegan Paul, New York.

Mokyr, J. (1990), *The Levers of Riches: Technological Creativity and Economic Progress,* Oxford University Press, New York.

Mutch, A. (2003), 'Communities of Practice and Habitus: A Critique', *Organization Studies,* vol. 24, no. 3, pp. 383-401.

Neuijen, B. (2001), 'Construction Research and Culture's Impact', In. *Culture in Construction: Part of the Deal? CIB Proceedings Publication 255,* W. Tijhuis (ed.), Tilburg University, Tilburg, Netherlands.

O'Grady, J., and Lampert, G. (1998), *The Underground Economy in Ontario's Construction Industry,* Ottawa, ARA Consulting Group.

Ontario Construction Secretariat, Peartree Solutions and Prism Economics and Analysis. (2003), *Aging Construction Workforce and Labour Market Shortages: Myth...or Reality?* Ontario Construction Secretariat [Cited Jan. 2003]. Available from: http://www.iciconstruction.com/site/pdf/AgingWorkforceJan2002.pdf.

Schein, E.H. (1992), *Organization Culture and Leadership,* 2nd ed., Jossey-Bass, San Francisco.

Seaden, G., Guolla, M., Doutriaux, J. and Nash, J. (2001), 'Analysis of the Survey on Innovation, Advanced Technologies and Practices in the Construction and Related Industries 1999', Statistics Canada, Science Innovation and Electronic Information Division, Statistics Canada, Ottawa.

Selmer, E. and de Leon, C. (1996), 'Parent Cultural Control through Organizational Acculturation', *Journal of Organizational Behaviour,* vol. 17, pp. 557-572.

Shapira, A. and Glascock, J. (1996), 'Culture of Using Mobile Cranes for Building Construction', *Journal of Construction Engineering and Management,* vol. 122, no. 4, pp. 298-307.

Shields, R. and West, K. (2003), 'Innovation in Clean Room Construction: A Case Study of Cooperation between Firms', *Construction Management and Economics,* vol. 21, no. 4, pp. 324-337.

Shove, E., Packwood, N. and Shields, R. (1997), *Factors Affecting Competitiveness in Construction,* Department of the Environment, London.

Slaughter, E.S. (1993), 'Builders as Sources of Construction Innovation', *Journal of Construction and Engineering Management-ASCE,* vol. 119, no. 3, pp. 532-549.

Sobel, D. (1995), 'From Grunt Work to No Work: The Impact of Technological Change on the Building Trades', In. *Re-Shaping Work: Union Responses to Technological Change,* C. Shenk and J. Anderson (eds.), Toronto, Ontario Federation of Labour, Technological Adjustment Research Programme, Our Times Publishing, Toronto

Sterne, J. (2003), 'Bourdieu, Technique and Technology', *Cultural Studies,* vol. 17, no. 3-4, pp. 367-389.

Taveb, M. (1991), 'The Effects of Culture on the Management of Organizations', In. *Competitive Advantage in Construction,* S. Male and R. Stocks (eds.), Toronto, Butterworth-Heinemann.

Trompenaars, A. and Hampden-Turner, C. (1998), *Riding the Waves of Culture,* The Economist Publications, London.

Wenger, E. (1999), *Communities of Practice,* Cambridge University Press, Cambridge.

Wilkinson, S. and Scofield, R.L. (2002), 'The Influence of the New Zealand Culture on Construction Activities', in Seymour, D. and Fellows, R. (eds.), *Perspectives on Culture in Construction,* CIB, Paris.

Winch, G. (1998), 'The Growth of Self-Employment in British Construction', *Construction Management and Economics,* vol. 16, no. 4, pp. 531-542.

Conclusion

A Roundtable on Construction Innovation – How to Make it Work?

André Manseau and Rob Shields
with Contributors and Invited Experts

This book started with the attempts of providing a comprehensive overview of managing innovation and changes in construction from a project, firm or industry level. We have rapidly realized that we were addressing a moving and changing field of analysis. The scope, the borders and the activities of construction are changing, as well as the concept of innovation in general – how we define innovation and what are the key processes to innovate.

In order to prepare this concluding chapter, we conveyed all authors of the book and other invited experts to a roundtable discussion in April 2003. We are very grateful to Frances Anderson, Sherif Barakat, Keith Hampson, Roger Miller, Frens Pries, George Seaden and Graham Winch who all joined both of us to the discussion. This exercise has been very fruitful and provided us with a number of key elements for identifying major recent changes and trends in construction and innovation, for stressing key lessons learned from the various attempts and approaches used for assessing and managing innovation in construction, for understanding main drivers and barriers, and for proposing directions to develop a better understanding of innovation and changes for practitioners, policy makers and researchers.

While we were final editing the book later in 2003, we reviewed the key learning from all chapters and integrate it with the result of our roundtable discussion. This chapter conveys the Editors' sense of the meeting and its conclusions.

Innovation is widely recognized as a key factor of competitivity and productivity of any economic activity. The development of new and significantly improved products or goods and/or the improvement of processes for producing and delivering these goods create value to both producers and customers. In the case of construction, a number of national panels identified the industry as a key to achieving wider social goals such as energy efficiency and necessary infrastructure for supporting the overall economy and quality of life. However, construction remains to a certain extent a craft industry, which appears to defy the rationale of innovation and its impact of wealth.

10.1 Major Recent Trends in Construction and Innovation

Innovation in construction has appeared to be not only driven by the 'improvement' rationale, but also stimulated by industry changes. Key changes in the business of construction have required innovations in products as well as in processes.

However, from the various trends identified in this work and recent events, it seams that the most important change in construction has not been stimulated by new products or technologies, but has occurred within the overall business behaviour with partners and customers. The most fundamental change appears to be in client behaviour. A number of studies have observed new contractual or procurement arrangements, an increasing importance of life cycle issues and trend toward a more services oriented industry (Chapters 2, 3 and 7). The logic has changed – from the traditional product delivered on pre-approved specifications to a life-long facility and its related services. Relationships between suppliers, contractors, facility managers and even some key-users are shifting toward long term, creating new games, coalitions and strategies.

The industry appears more complex and broader, no longer limited to contractors. Indeed, the industry growth is not so much in term of its overall economic activities, but in the increased interactions and merging activities within related sectors. Some building product manufacturers are becoming installers as well as provide after sale services, some engineering firms or contractors are becoming operators of large facilities, and large facility owners have their own in-house design and building capacity.

The increased complexity of interrelations has opened new needs and opportunities. New organizational changes, social interplays and governance frameworks have provided firms with significant competitive advantage (Chapter 3, 5 and 6). In the case of large projects (particularly discussed in Chapter 6), there have been significant investments in developing the front-end of projects (social pressures, interplays, and a framework of governance for providing the users and key stakeholders with the required flexibility for shaping the product).

Within this new complex arena, innovation brokers are becoming more important by facilitating interactions between numerous different players (Chapter 5). The use of sophisticated Information and Communication Technologies (ICT) is also facilitating management of complex interactions.

This change in the logic of business has impacted in the need for new performance measures for addressing social issues, safety and services. The importance effective interactions and even strategies between players have motivated the need for developing new industry policies and action plans.

Performance assessment is still difficult to define and construction activities are still in a large part 'as usual', but there is definitely a shift toward social and organizational issues. The management culture, as well as the various 'habitus', are perhaps the last frontier or resistance to change (Chapter 8 and 9). Nevertheless, the construction hard-and-rough is slowly becoming soft-and-wise.

10.2 Key Lessons Learned from the Various Attempts

The use of existing indicators for measuring innovation, particularly developed in the context of the manufacturing sector, has appeared limited in construction (Chapter 4). Opened up to a larger definition of innovation, this work has shown some very interesting results. However, measures of organizational innovation, flows and diffusion of innovations are still in their infancy.

Statistical information relies on surveys that are sent to firms and a representative sample of firms can provide statistical picture of the whole industry. For decades, national statistics agencies define the construction industry as constituted of the firms working directly on the construction site, i.e. prime and trade contractors. Well-established methodologies and definitions support comparative analysis between countries as well as long time period studies.

However, a number of studies have challenged this approach of defining and gathering information on construction activities. There are two major misleading from this approach. The first one is that key construction activities, and often the most innovative ones, are conducted off-site, such as construction design and planning and product manufacturing. Secondly, construction activity is essentially project-based, involving multi-partners, and not well captured when using a firm based approach as the level of analysis.

A system approach has been recently developed for measuring innovation in the construction sector where the focus is no longer limited to a single firm but on the process of innovation from its generation, transmission and use within a network of players. Still few studies have used this new approach, but again surveys were designed for firms, using available databases. Nevertheless, these studies stressed the importance of business practices and collaborative arrangements to succeed – a more people centred that technical solution to problems. However, the measure of diffusion of innovations, from their generation to use, poses serious challenges, as an innovation is often re-invented or adapted by each player.

A few case studies have been conducted, but they are still limited and provide limited weight for an industry wide perspective. However, these studies have been very helpful in identifying new critical factors for success, such as the importance of influencing institutional framework and of shaping client relationship. But many challenges still persist on how to measure that? Much more empirical evidences are needed.

10.3 Main Drivers and Barriers of Innovation

The social component of innovation is the most important issue. Clients, and particularly the government that is still a large client, are key drivers of changes. For instance, new forms of procurements have been introduced in many countries; UK, North America, France, Australia. However, changes required a culture shift, and it appears that in some other countries (Chapter 8), new procurement arrangements have failed.

However, the majority of the industry is moving very slowly. The construction industry has not been able to accomplish innovate to nearly the same degree as other sectors. Some large multi-national corporations have been able to advance innovation significantly, as well as a few small 'smart' firms have gained international and high-profit market niches in other sectors. Indeed, innovation appears to be strongly correlated to the size of firms and to their international focus. Unfortunately, there are very few such firms - the construction industry being highly fragmented, constituted of a number of very small firms.

Increasingly, clients want to be part of the process, as they cannot define 100 per cent of needs prior to start. A building, even a simple house, is a complex product which will fulfill many different needs, some expected at the beginning but a number that will continuously emerge as the user is realizing all of the untapped potentials. However, construction professional appears to have difficulties in addressing that need, still working with the traditional professional paradigm – 'tell me want you want, then let me do it'. Larger clients have their own in-house experts managing the overall construction process, from the design stage, to building and maintenance. They usually contract-out the less creative or innovative part of the work – keeping a tight control on the design and pre-defining themselves most of the specifications. There are a certain distance between the professional and the client – interactions are limited in scope and in time.

A more pro-active and progressive (i.e. learning from each other) approach is a very fundamental change in the way of doing businesses and in the client relationship – some are taking of a change in governance, others as a change in culture.

Not all clients and/or products required new interactions, but a wealthy industry is one that addresses changes and makes the overall system work with success. There is an impact of individual leaders who transcend traditional boundaries and there is certainly a key role for governments, as major client as well as facilitator of macro changes such as culture, institutions and systems. National visions or action plans, sponsored by governments, have had a significant impact in many countries (UK, USA, Australia).

10.4 Implications for Future Directions

With regard to measures of innovation and changes, perhaps one of the main current limitations is the focus on physical building or facility as 'final' product.

Delivery systems should be much more comprehensive, as half the construction industry in many European countries consists of repair and maintenance, and the increasing importance of partnering arrangements including operations and maintenance of facilities. Another significant challenge in measuring innovation is to take into consideration its diffusion – that always involves re-innovation.

A movement toward innovation and change in construction would require a broader perspective than technology – from policy, to process to products: culture and social governance, procurement systems and strategies, reach/exceed clients' expectations on new performance criteria beyond initial cost (such as flexibility of use and life cycle cost). This broad perspective is crucial for any action plan and has many implications for researchers, practitioners and policy makers.

From the research point of view, multi-level studies – project, firm, institutional set-up; with an opening to policy, business and 'soft' issues would be required. Do we have a development of industry clusters – network of firms with suppliers and large firms – operating in the international market, such like in high-tech sectors? Perhaps some clusters are emerging in some engineering areas, water utilities, office spaces, or in modular residential building or components? These should be investigated.

For construction practitioners, building tomorrow would mean moving above traditional boundaries in the types of activities as well as in time scaling. Less focus on technology and products, and more on services; less protective of a role, but establishing new rules and partnering arrangements.

Finally, construction action plans and programs would imply for policy makers an active role in mobilizing key players, formulating strategies, and partnering with the private sector and all other stakeholders for developing sustainable solutions. How to facilitate and support change in such a complex and large industry? A shift from regulator and lowest-bid procurement to a facilitator and partner for continuously changing collective values is the key challenge for policy makers and government role.

Index

Advanced management techniques 68
Advanced practices 43, 61, 68-69, 94
Advanced technologies 68-70, 76-77

Behaviour iii, 2, 6, 50, 78, 94, 107, 126, 158-159, 172, 176
Builders 1, 8, 17
Buildings 1-2, 6, 8, 10, 12, 19, 45, 57, 61-62, 89, 123-124, 129, 134, 140, 163
Business 16, 20, 43, 49, 52-53, 60, 78, 86, 101,111, 117, 119, 123, 130, 136, 141, 146, 149, 153, 160, 169
  creative futures 63, 82
  practice innovations 75
  short-term orientation 149
  strategic plan 4, 46, 74
  systems 46-48, 92

Challenges 2, 3, 10, 96, 103, 107, 115, 129, 133, 134, 140, 155, 157, 171
Champions 3, 13, 17, 18, 124. *See also* Leader
Clients as drivers of change iii, iv, 3-4, 6, 8-9, 14-18, 45-46, 50-52, 54, 69, 73-76, 83, 85-86, 93-95, 97 111, 116, 118, 129, 132, 158 162-163, 171, 176, 178
Clusters 43, 46
Commercialization. *See* Discovery
Communication 45, 48, 52, 57, 67-69, 75, 131-132, 147
Companies 3, 13, 61, 64, 74-75, 97, 128-129, 142, 144-145, 149-150, 152, 154
Competition iii-iv, 7, 47, 49, 53-54, 82, 84, 94, 128, 136, 140, 175
Complex product systems (CoPS) 2, 5, 48, 82
Complex systems industry 16, 43, 82, 85, 87, 89, 98
Complexity 19, 50-51, 58, 118, 130 146, 176
Consensus decision making 14, 49, 107, 128, 135, 154
Consolidations 11, 49-50, 133, 157
Construction iv, 1-3, 6, 9-20, 43-45, 48, 50-52, 57-60, 70-80, 82-83, 85-87, 89, 91-92, 94, 96-97, 101, 104-105, 111, 113, 119, 124-125, 127-136, 139-140, 144-145, 148-153, 157, 159-160, 154-171, 175-177, 179
  activities 2, 11-12, 15, 20, 47-48, 50-51, 53, 94, 123, 166, 177
  cluster 2, 57-60, 63, 77, 179
  industry iii-iv, 2-3, 5, 6, 10, 16-17, 43-48, 50-52, 57-60, 68-69, 75-76, 81-83, 86-88, 90, 93-94, 111, 123-124, 128-133, 139-143, 145-146, 149-151, 153-154, 158, 165, 167, 170-171, 177-178
Construction productivity 2
Contracting 2, 13, 47, 52, 57-58, 108
Cooperation iii, 7, 75, 130-131, 141, 144, 160
Cooperative technical organizations 87
Creative work 64
Creativity: approach 115
Culture iv, 3-4, 11, 52, 77-78, 86, 133, 135, 157-161, 163, 165, 167-172, 177, 179. *See also* Organizational Culture

'Decisioneering' 108-109
Decision making 52, 106, 116

Designers 1, 8, 16-17, 88, 123, 136, 166-167, 170
Discovery 107, 148
  Commercialization 9, 129-130
Dissemination, skill 6, 127
Distributed innovation processes (DIPS) 14, 84
Doubt 119

Ecology 82-83
Economic environment 19, 51
Economy, and construction 25
  Turbulence 106. *See also* Risk
Empiricism 1. *See also* Thinking
Evolution 1, 49, 52, 67, 82-83, 94, 106, 126-127, 129
Experience 6, 9, 13, 17-19, 50, 87, 92, 102, 107, 115, 119, 147, 149, 151-153, 157, 162, 168, 170-171
Expertise 48-49, 59, 62, 94, 126, 136, 172

'Firm as Actor' approach 57, 60, 78
Flows 11, 14, 44, 51, 59, 68, 72-74, 79, 113, 116, 120, 136, 163, 177

Globalization 49, 86, 125, 127, 131, 135
Goods, construction 125
Golden period 51
Government, intervention 135

Human resources 73, 158
  'manpower' 79

Industrial Revolution 46, 82
Information iii, 4, 14-16, 18, 57, 61, 67-68, 72-75, 79, 85, 91, 110-111, 125, 162, 168-170
  flow of 14-16, 72-75, 79
Innovation iii, 1-2, 5, 7-8, 10, 12, 15-20, 43-45, 50-53, 57, 64, 66, 81-84, 87, 89-93, 95, 97, 114, 118, 126-127, 129-131, 132, 134, 139-146, 154-155, 157, 160, 162-163, 165, 167, 169, 171, 175, 177-179
  architectural or configurational 11, 17

barriers or impediments to 1, 4, 6, 8, 17-20, 45-46, 83, 140, 157-158, 167-168, 170-171, 175-176
basic parameters 14, 59-60, 65
broker or knowledge broker 3, 16-18, 51, 82-83, 86-92, 94-95, 97, 128
cooperation 7, 9
determinants of 2, 13, 17, 44, 74, 90
diffusion of iii, 2, 8-9, 49, 92, 94-96
drivers 3, 43, 50, 81
fostering iv, 7, 15, 19, 45-46, 52-53, 72, 74, 82-84, 95, 127-128, 130, 133, 135-136, 141, 143, 154
games of 3, 84, 86, 97, 176
ideal environment 46
importance 9, 72, 74, 163
incremental 1, 10, 12-13, 95, 141, 144-146
individualism 49
interaction with basic research 5
interdependent nature 6-7, 11, 16, 45, 47, 50-51, 64, 68, 86, 128
invention versus 9
management of iii-iv, 2-3, 81-82, 85, 94, 129-130, 147
modular (modes) or product 10-12, 43-44, 49, 52, 66, 92-94, 142, 145, 170
process 7, 9-11, 13-14, 16-17, 47-51, 81-82, 84-88, 92, 94-95, 97, 130, 142, 145, 148, 157-158, 160-161, 165-166, 170
productivity growth 9
radical 11-12, 84, 144
social 20, 64, 126, 131, 157
social structure 50, 86
standardization 12
strategies for enhancing 4, 45, 61, 71
structure 87-88, 91
system 11-13, 17-18, 43, 45, 57, 60-61, 74, 83-84, 86-87, 91-92, 95, 127
thrives in interactive environment 11, 18, 46-47
ultimate goal 13

Innovation/diffusion model 2, 95
Innovations: as a competitive advantage
    iii, 7, 52, 84, 94, 124, 129,
    135-136, 146, 171, 176
Innovator 6-7, 14, 60-61, 63, 66, 74,
    76-77, 79, 84, 91, 94, 96, 141,
    145, 154, 161, 164-165, 171
Invention 5, 9, 14, 64, 95, 147, 167

Know-how 61, 64
Knowledge 6, 9, 10, 45, 46, 64, 72-75,
    78-79, 92-93, 109, 124,
    147-148, 151, 160-161, 163,
    172
  dynamics 16, 46, 50-51, 67-68, 96, 130
  nature of 12, 124, 136
  uncertainty 95
Knowledge flows 44, 72, 74, 90, 136
Knowledge gain 8
Knowledge networks 12, 45-46, 50-51

Labour 3, 45, 47, 124, 127-128, 130,
    157, 162, 164-172
Leader or leadership 15, 49, 118-119,
    129-130, 133, 135-136, 148,
    161
Learning 11, 15, 18, 49-50, 83, 90, 94,
    96, 126, 135-136, 149, 160,
    162, 165
Linearity 14, 43-44, 95
Low morale 3

Management iii, 13-15, 48, 58, 73, 83,
    85, 88, 93, 109, 131-132,
    139-140, 147, 149-155, 158,
    161, 163, 166, 169, 171, 176
  cooperative strategies 10, 48
  methods 2, 5, 68-69, 101-133, 141
  of risks 11, 104, 108, 114, 118
  strategic 3, 13, 47, 83, 140, 147, 158
  strength of 15
  study of 1, 13, 157, 159, 161
Market pull model 44, 52, 131
Mediators 18
Models for knowledge 14, 43, 96, 111,
    117, 147, 154, 169

National construction business system
    14, 91-92, 132-133, 135
Neo-Liberalism 123-125
  'laissez-faire' 123-125
Network thinking 43, 45-46, 49-51, 73,
    83-86, 89, 91-92, 97, 110-111
    115-116, 126-127, 133, 177
New products 44, 50, 52, 60, 73-74, 129,
    134, 161, 167, 176
  generating 6, 9-11, 16, 46-47, 49, 52,
    73
  inefficiency of 73, 87
Nonlinearity 43
North American Industry Classification
    System (NAICS) 57

Organizational culture 3, 158, 160-161.
    See also Culture
Oslo model or OECD 1-2, 9, 43-44,
    46-47, 60, 68-69, 72, 74, 77-78
    123, 127, 129, 133, 141,
    167-168

Paradigms of organization management
    141, 146-149, 153-154, 178
Patent 43, 64, 85, 125, 140
Perspectivalism 12-13
Pragmatism. *See* Creativity
Preparation 12, 47, 164
  critical aspects 58, 61, 64
Problem-solving efforts 9, 85, 94, 160,
    167, 171
Productivity 2, 7, 9, 46-47, 139, 145,
    157, 164, 166, 168-169, 171,
    175. *See also*
    Construction productivity
Production systems 43, 46-47, 58-59, 72,
    132
Products 1, 7, 10, 12, 14-15, 17, 20,
    44-45, 48, 51, 53, 57, 59,
    61-64, 66, 68-69, 72-74, 77-
    79, 82, 85, 87, 89, 96, 112,
    123-124, 128, 130, 136,
    139-140, 146, 162, 166,
    169-171, 176

Rationalism. *See* Thinking
Reductionism. *See* Thinking
Relationships:

emotional 46, 49, 50, 52, 75, 77, 86, 93, 119, 158, 162, 166
Research 18, 59, 62, 102, 110, 117, 119 128-129, 131-132, 159, 161, 163-164
  basic iii, 10, 44
  bias 1
  codification and backwater 13
  empirical 13, 44
  industrial 7, 19, 85-87
  interaction with innovation 9, 12-13, 44, 60, 70
  literature 1, 5, 19-20, 81-83, 84, 86-87, 90, 92, 96-97
  nature of 7
  politics 13-14, 20
  strategic 91
  targeted 3, 75-76, 92-93
Rewards 18, 112
Risk 2, 6, 11, 15-19, 45, 50, 73, 79, 101-120, 130, 132, 135, 146, 159, 162, 170

Schumpeter, J. 5, 82, 84, 123, 145
Science 1, 85, 103, 147
  application technologies 44, 50, 60, 67, 69, 74
Scientific approach 44, 92, 147, 149, 158
  need for constant monitoring
Scientific method. *See* Science; Thinking
Serendipity. *See* Discovery
Skills 6, 11, 17, 43, 47, 57, 78, 91-92, 98, 111, 136, 150-153, 157-158, 163-172
  'deskilling' 166
Social capital 3, 86, 126, 133, 160. *See also* Human Resources; Labour
Social networks 9-10, 49, 126, 162
Social process 19, 43, 49, 52
Social values 133, 136
Standards 2, 18-19, 49, 52, 83, 85, 91, 94, 123, 125-127, 129, 133, 135-136, 145, 154
Standardization 2, 5, 63, 96, 169
Strategic alliance, *See* Cooperation
Systems 2, 5, 10-12, 16-18, 43, 46, 48, 50, 62, 69, 72-73, 75, 78, 82-89, 91-92, 102, 108-110, 113-114, 116, 123, 125-127, 129-131, 135, 143, 154
Systems approach 57, 67, 72, 78-79, 177

Technique 5, 8-9, 77, 83-84, 92, 94, 108, 165, 172
Technology iii-iv, 3, 10, 43-47, 49, 60, 67, 75, 84-85, 92, 107, 129, 133, 136, 139, 146, 148, 162, 166-167, 171, 176, 179
  diffusion of 44, 46, 50, 52, 78, 129
  essential knowledge 89, 148
  fusion 87, 90
  government and industry control 86, 102, 128
  modern 20
  nature of 50, 129
  technological market surveys 68, 77
Technology push or science-based model iii, 1, 43-44, 52, 131
Technoscientific age
  rapidity of change 8
Thinking iii, 13
  Time 5-9, 11, 16, 48, 83-84, 90, 94, 113, 116-118, 164-165, 167-169
Tit-for-tat. *See* Management
Traditional or management 101, 108
Training 64, 73, 128, 133, 149, 157-158, 163, 167-169, 171-172
Trust 6, 48, 50, 77, 86, 91, 111, 127, 133-134, 136, 154, 158
Turnover rate and morale 3, 162

Work 3, 47, 57-58, 61-62, 64, 75, 132, 154, 157, 163-166, 168-171
  models for knowledge organizations
  public works 123, 128, 132
  process 6, 47, 115, 126, 162-163
  relationships 6, 9, 14, 46-47, 51, 94, 107, 126, 135, 160, 167
  sectorization 96
Work bounds: integration 73
Work environment 6, 46

Zero-based thinking 18